Fortran 90
for Engineers

Delores M. Etter
Department of Electrical and Computer Engineering
University of Colorado, Boulder

The Benjamin/Cummings Publishing Company, Inc.

Redwood City, California · Menlo Park, California
Reading, Massachusetts · New York · Don Mills, Ontario
Wokingham, U.K. · Amsterdam · Bonn · Singapore
Tokyo · Madrid · San Juan

Photo Credits
Chapter 1: Courtesy of NASA.
Chapter 2: © David Madison.
Chapter 3: © Jim Pickerell/TS Images.
Chapter 4: © W. Metzen/AllStock.
Chapter 5: Courtesy of NASA/Ames.
Chapter 6: © Dan Suzio.
Chapter 7: Richard Wagner/Benjamin Cummings.
Chapter 8: © Ray Richardson/Animals Animals.

ISBN 0-8053-6464-1

The Benjamin/Cummings Publishing Company, Inc.
390 Bridge Parkway
Redwood City, CA 94065

Acknowledgements
The author and publisher gratefully acknowledge the contributions of the following individuals who reviewed the manuscript: John Lilley, University of New Mexico; Alan Genz, Washington State University; Howard Silver, Fairleigh Dickinson University; Howard Miller, Virginia Western Community College; Nigel Wright, Leeds University, England; Brad Nickerson, University of New Brunswick, Canada; Peter Kattan, Louisiana State University; Craig T. Dedo, Elmbrook Computer Services; Lt. Cmdr. Joseph Bradley, United States Navy.

The author and publisher would also like to thank Guy Ceragioli of Lahey Computer Systems who supplied a Fortran 90 compiler for the development of this module.

Contents

1 Solving Problems with Fortran 90

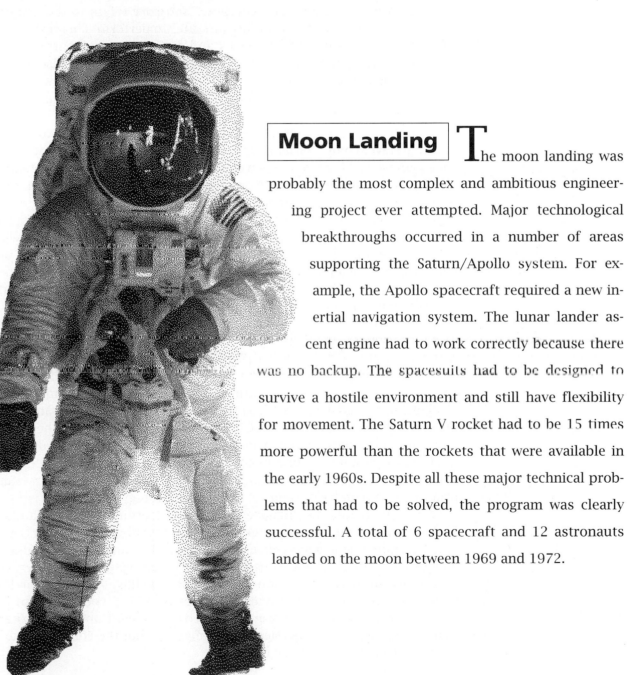

Moon Landing The moon landing was probably the most complex and ambitious engineering project ever attempted. Major technological breakthroughs occurred in a number of areas supporting the Saturn/Apollo system. For example, the Apollo spacecraft required a new inertial navigation system. The lunar lander ascent engine had to work correctly because there was no backup. The spacesuits had to be designed to survive a hostile environment and still have flexibility for movement. The Saturn V rocket had to be 15 times more powerful than the rockets that were available in the early 1960s. Despite all these major technical problems that had to be solved, the program was clearly successful. A total of 6 spacecraft and 12 astronauts landed on the moon between 1969 and 1972.

INTRODUCTION

This chapter introduces you to problem solving by first examining the class of problems *Fortran 90* was designed to solve. The chapter then presents a five-step process that can be applied to solving a wide variety of problems and gives a simple example that uses this five-step process to compute the average of a set of laboratory measurements.

1-1 FORTRAN 90

A *computer* is a machine designed to perform operations that are specified with a set of instructions called a *program*. Computer *hardware* refers to the computer equipment, and computer *software* refers to the programs that describe the steps we want the computer to perform. This software can be written in a variety of languages and for a variety of purposes. This section describes the Fortran 90 language; the next section discusses the problem-solving process we will use to write Fortran 90 programs.

High-level languages are computer languages that have English-like commands and instructions. Originally developed in the 1950s for technical applications, FORTRAN (short for FORmula TRANslation) was one of the first high-level computer languages. Since FORTRAN's creation, its power, speed, and ease of use have been consistently enhanced with new features. FORTRAN 77 is a version based on standards accepted in 1978; Fortran 90 is the version based on standards developed over several years and accepted in 1991. The new Fortran 90 standard capitalizes only the first letter in Fortran to distinguish it from earlier versions. Today Fortran is a widely used programming language for solving scientific and engineering problems because of its ability to perform mathematical calculations quickly and efficiently. C is another programming language frequently used to solve engineering and scientific problems.

Alternatives to writing a program in a high-level language exist, such as using specialized software tools and applications packages. Prewritten software routines designed to solve common scientific and engineering problems allow you to avoid writing a program that has already been created. Powerful engineering, mathematical, and statistical functions are available in spreadsheets, such as Lotus 1-2-3 and Quattro Pro, and mathematical packages, such as MathCAD and MATLAB. However, when pre-existing software tools are not available on your computer, or you don't have the time to learn how to use an application package, or the problem you're working on is too large or complex for an application package, writing a program in a high-level language is the best strategy.

Learning a high-level language is similar to learning a foreign language. Each step you want the computer to perform must be described in the specific computer language syntax, or grammar. Fortunately, computer languages have small vocabularies and no verb conjugations; however, computers are unforgiving in punctuation and spelling. A comma or letter incorrectly placed will usually cause errors that keep your program from working properly. A special program called a *compiler* translates a program written in a high-level language into a language that the computer can understand called machine language.

1-2   A FIVE-STEP PROBLEM-SOLVING PROCESS

Problem solving is an activity in which we participate every day. Problems range from analyzing chemistry lab data to finding the quickest route to work. Computers can solve many of our problems if we learn how to communicate with them in computer languages such as Fortran 90. Some people believe that if they describe a problem to a computer, it will solve it for them. Programming would be simpler if this were the case; unfortunately, it is not. Computers can perform only the steps that you specify in detail. You may wonder then, why go to the effort of writing computer programs to solve problems if every step must be described carefully? The answer is that computers can perform the tasks extremely accurately and with fantastic speed (millions of arithmetic operations per second). In addition, computers never get bored. Imagine sitting at a desk analyzing laboratory data for 8 hours a day, 5 days a week, year after year. Yet thousands of laboratory results must be analyzed every day. Once the steps involved in performing a particular analysis (such as computing an average) have been carefully described to a computer, it can analyze data 24 hours a day, with more speed and accuracy than a group of technicians.

Much of this module will focus on teaching you the Fortran 90 language, but any computer language is useless unless you can break a problem into steps that a computer can perform. The approach we use to write Fortran programs is based on the following five-step problem-solving procedure. A general procedure is shown on the left, and the procedure we will use to develop programs is shown on the right.

A General Problem-Solving Model	Fortran Problem-Solving Model
1. Define the problem.	State the problem clearly.
2. Gather information.	Describe the input and output information.
3. Generate and evaluate potential solutions.	Work the problem by hand (or with a calculator) for a simple set of data.
4. Refine and implement a solution.	Develop a solution that is general in nature.
5. Verify and test the solution.	Test the solution with a variety of data sets.

We now illustrate this five-step problem-solving process using the familiar problem of computing an average of a set of five data values from a lab experiment.

Step 1 defines the problem. This often involves restating the problem in a clear unambiguous manner. For this example the problem statement is as follows:

Compute the average of a set of five experimental data values.

Step 2 consists of gathering information. We need to describe carefully any information or data needed to solve the problem and then identify the values to be computed. These items represent the input and output for the problem; collectively they are called *input/output (I/O)*. The I/O description in this example is the following:

Input—the list of experimental data values
Output—the average of the data values

Step 3 generates and evaluates potential solutions. During this step we need to work the problem by hand or with a calculator, using a simple set of data. The following example computes the average of a set of values:

Lab Measurements—4/23/95

Number	Value
1	23.43
2	37.43
3	34.91
4	28.37
5	30.62
	154.76

Average = sum/5 = 30.95

For a simple problem such as this, there may be an obvious solution, and identifying it may take only a few minutes. More complex problems may be solved in several ways, and determining the most suitable solution can be a difficult and time-consuming task.

Step 4 refines and implements a solution. In this step we describe, in general terms, the operations we performed by hand. The sequence of operations that solves the problem is called an *algorithm*. The procedure that we use in algorithm development is called *top-down design*, because we start at the top with the original problem and break it down into smaller problems that can be addressed separately. Top-down design uses two techniques: *decomposition* and *stepwise refinement*. Decomposition is a form of "divide and conquer," in which you identify the pieces of the problem that must be solved sequentially. Stepwise refinement successively refines each smaller piece of the solution using more detail. The refining continues until the solution is specific enough to convert into computer instructions.

Top-down design has several advantages. First, it helps us write programs that are easier to understand, because the instructions in the computer program will follow the initial decomposition to emphasize that we are breaking the solution into a series of sequentially executed steps. To assist in refining the decomposition into a more specific solution, we use flowcharts, which show the steps in an algorithm in graphic form, or pseudocode, which presents algorithm steps in a series of English-like statements. Chapter 3 presents the details of flowcharts and pseudocode.

Decomposition

Read and sum the 5 data values.
Divide the sum by 5.
Print the average.

The refinement of the decomposition for this example uses pseudocode to specify the details or to outline the steps in the algorithm.

Pseudocode

1. Set the sum of the values to zero.
2. Set a count of the values to zero.
3. As long as the count is less than 5,
     Read the next data value,
     Add it to the sum,
     Add 1 to the count.
4. Divide the sum by 5 to get the average.
5. Print the average.

Once the detailed algorithm is described, we are ready to convert it into Fortran 90. The statements for solving this problem will not all be presented until Chapter 3, so there are items in the solution that you probably won't understand at this time. You can, however, see how similar it is to the decomposition and pseudocode.

Fortran Program

```
!------------------------------------------------------------------!
program compute_average
!
!   This program computes and prints the average
!   of a set of five experimental data values.
!
!
!      Define and initialize variables.
       implicit none
       integer :: count=0
       real :: average, sum=0.0, x
!
!      Prompt the user to enter the data values.
       print*, "Enter five data values, one per line:"
!
!      Read and sum the data values.
       do while (count < 5)
          read*, x
          sum = sum + x
          count = count + 1
       end do
!
!      Compute average.
       average = sum/5.0
```

```
!
!    Print average.
     print 10, average
  10 format(1x,"The average is ",f5.2)
!
end program compute__average
!-------------------------------------------------------------!
```

Step 5 in the problem-solving process verifies and tests the solution. Compiling the program verifies that the syntax is correct. We then test the program with many different sets of data to be sure the logic is correct, and this is more difficult than testing for correct syntax. A correct algorithm to average data should work properly for any valid set of data. If the data from our hand example were used in the Fortran program presented here, the output on the terminal screen would be the following:

```
The average is 30.95
```

Algorithm testing is discussed further in later chapters.

The five-step problem-solving process is demonstrated throughout this module in applications. The disciplines of the applications are as follows:

Applications	Across the Disciplines	
Application	**Discipline**	**Chapter**
Bacteria Growth	Biomedical Engineering	2
Stride Estimation	Mechanical Engineering	2
Light Pipes	Optical Engineering	3
Rocket Trajectory	Aerospace Engineering	3
Timber Regrowth	Environmental Engineering	3
Critical Path Analysis	Manufacturing Engineering	4
Sonar Signals	Acoustical Engineering	4
Wind Tunnels	Aerospace Engineering	5
Power Plant Data Analysis	Power Engineering	5
Oil Well Production	Petroleum Engineering	6
Simulation Data	Electrical Engineering	6
Protein Molecular Weights	Genetic Engineering	7
Migration Paths	Environmental Engineering	8

SUMMARY

This chapter described Fortran 90 as a language developed to solve engineering and scientific problems easily and efficiently. Because engineering involves problem solving, it is important to begin your investigation of Fortran with a solid methodology for solving problems and then to learn how to use the computer to help solve the problems you encounter. We discussed the fundamental concepts of computers and computing and the process for converting a problem solution into a form the computer can understand and execute. The following five-step procedure for developing problem solutions (algorithms) was presented:

1. State the problem clearly (problem statement).
2. Describe the input and output information (input/output description).
3. Work the problem by hand (or with a calculator) for a simple set of data (hand example).
4. Develop a solution that is general in nature (algorithm development).
5. Test the solution with a variety of data sets (testing).

The solution is developed using top-down design. Decomposition assists us in describing the general steps that have to be performed to solve the problem. Stepwise refinement guides us in refining the steps and adding necessary details. We then use Fortran 90 to implement these solutions.

Key Words

algorithm	high-level language
compiler	input/output (I/O)
computer	program
decomposition	software
Fortran 90	stepwise refinement
hardware	top-down design

References

For further reading on the 10 top engineering and scientific achievements that were discussed in the overview at the beginning of this text, we recommend the following references. Many of these references are from the National Academy of Engineering brochure entitled "10 Outstanding Achievements 1964–1989," published in 1989.

MOON LANDING

Hess, Wilmot, et. al. "The Exploration of the Moon." *Scientific American,* October 1969, pp. 55–72.
Stix, Gary, ed. "Moon Lander." *Spectrum,* Vol. 25, No. 11, 1988, pp. 76–82.

APPLICATION SATELLITES

Canby, Thomas Y. "Satellites That Serve Us." *National Geographic,* September 1983, pp. 281–334.

Heckman, Joanne. "Ready, Set, GOES: Weather Eyes for the 21st Century." *Space World,* July 1987, pp. 23–26.

MICROPROCESSORS

Garetz, Mark. "Evolution of the Microprocessor: An Informal History." *BYTE,* September 1985, pp. 209–215.
Greenberg, Donald P. "Computers and Architecture." *Scientific American,* February 1991, pp. 104–109.

COMPUTER-AIDED DESIGN AND MANUFACTURING

Loeffelholz, Suzanne. "CAD/CAM Comes of Age." *Financial World,* October 18, 1988, pp. 38–40.
Mitchell, Larry D. "Computer-Aided Design and Manufacturing." *McGraw-Hill Encyclopedia of Science & Technology.* New York: McGraw-Hill, 1987.

CAT SCAN

Andreasen, Nancy C. "Brain Imaging: Applications in Psychiatry." *Science,* March 18, 1988, pp. 1381–1388.
Sochurek, Howard. "Medicine's New Vision." *National Geographic,* January 1987, pp. 2–40.

ADVANCED COMPOSITE MATERIALS

Chou, Tsu-Wei, Roy L. McCullough, and R. Byron Pipes. "Composites." *Scientific American,* October 1986, pp. 192–203.
Steinberg, Morris A. "Materials for Aerospace." *Scientific American,* October 1986, pp. 67–72.

JUMBO JETS

Ingells, Douglas J. *747: Story of the Boeing Super Jet.* Fallbrook, CA: Aero Publishers, 1970.
Stewart, Stanley. *Flying the Big Jets.* New York: Arco Publishing, 1985.

LASERS

Berns, Michael W. "Laser Surgery." *Scientific American,* June 1991, pp. 84–90.
Jewell, Jack L., James P. Harbison, and Axel Scherer. "Microlasers." *Scientific American,* November 1991, pp. 86–94.

FIBER-OPTIC COMMUNICATIONS

Desurvire, Emmanuelf. "Lightwave Communications: The Fifth Generation." *Scientific American,* January 1992, pp. 114–121.
Drexhage, Martin G. and Cornelius T. Moynihan. "Infrared Optical Fibers." *Scientific American,* November 1988, pp. 110–116.

GENETICALLY ENGINEERED PRODUCTS

Barton, John H. "Patenting Life." *Scientific American,* March 1991, pp. 40–46.
Bugg, Charles E., William M. Carson, and John A. Montgomery. "Drugs by Design." *Scientific American,* December 1993, pp. 92–101.

2 Arithmetic Computations and Simple Programs

Stride Estimation The motion of a simple pendulum can be used to model the motion of a leg in a natural stride. Using this model, it can be shown that the time that it takes for a stride is proportional to the length of the leg. Thus, the longer the leg, the longer the stride. If we know the length of a stride and the time it takes to make a stride, we can compute the walking speed, which is the number of feet per second traveled while walking. As you might have guessed, people with longer legs have faster walking speeds and go farther per second. These results do not apply to running, because the leg does not swing freely in a running gait, and thus it cannot be modeled by a simple pendulum. With an accurate model of a leg, new prosthetic legs can be designed using advanced composite materials that are lightweight and strong.

INTRODUCTION

Arithmetic operations (adding, subtracting, multiplying, and dividing) are the most fundamental operations performed by computers. Engineers and scientists also need other routine operations, such as raising a number to a power, taking the logarithm of a number, or computing the sine of an angle. This chapter discusses methods of storing data with Fortran 90 and develops the statements for performing arithmetic calculations with that data. We also introduce statements for simple data input and output. With this group of statements, we can write complete Fortran 90 programs.

2-1 CONSTANTS AND VARIABLES

Numbers are introduced into a computer program either directly with the use of constants or indirectly with the use of variables. *Constants* are numbers used in Fortran 90 statements, such as -7, 32.0, and 3.141593. Constants may contain plus or minus signs and decimal points, but they may not contain commas. Thus, 3147.6 is a valid Fortran constant, but 3,147.6 is not. Constants are stored in memory locations but can be accessed only by using the constant value itself.

A *variable* represents a memory location that is assigned a name. The memory location contains a value; we reference that value with the name assigned to that memory location. We can visualize variables, their names, and their values as shown:

amount	36.84	volume	183.0
rate	0.065	total	486.5
temperature	17.5	info	72

Each memory location to be used in a program is given a name and may be assigned a value using a Fortran statement. For example, the memory location named rate has been assigned the value 0.065 in the previous diagram.

Variable Names

Each variable must have a different name, which you provide in your program. The names may contain 1 to 31 characters consisting of alphabetic characters, digits, and the underscore character (__); however, the first character of a name cannot be a digit. Fortran does not distinguish between uppercase and lowercase characters in variable names. In this module we will use lowercase characters in variable names. The following are examples of both valid and invalid variable names:

Variable Name	Valid or Invalid
distance	Valid
time	Valid
pi	Valid
$	Invalid — illegal character ($)
tax-rate	Invalid — illegal character (−)
tax__rate	Valid
B1	Valid
2X	Invalid — first character is a digit
quadratic__equation__solution	Valid

Data Types

Fortran 90 contains five different types of values: integers, real values (both single-precision and double-precision), complex values, character values, and logical values. A discussion of these data types is given after the following list of examples:

Data Type	Examples
Integers	32, − 7
Single-precision values	− 15.45, 0.004
Double-precision values	3.1415926536D0, 1.000000006D0
Complex values	1 − 2i, 5i (where i = $\sqrt{-1}$)
Character values	"velocity", "Report 3"
Logical values	.true., .false.

Integers, real values (single-precision and double-precision), and complex values represent numerical values. This chapter focuses on integers and single-precision real values; double-precision and complex values are discussed in Chapter 7. Logical values are discussed in Chapter 3; character values are discussed in Chapter 7.

An *integer value* has no fractional portion and no decimal point, such as 16, − 7, 186, and 0. On the other hand, a *real value* contains a decimal point and may or may not have digits past the decimal point, such as 13.86, 13., 0.0076, − 14.1, 36.0, and − 3.1; real values are also called floating-point values.

A memory location can contain only one type of value. The type of value stored in a variable is specified by two methods: implicit typing or explicit typing. With *implicit typing,* the first letter of a variable name determines the value type that can be stored in it. Variable names beginning with the letters *i, j, k, l, m,* or *n* are implicitly used to store integers. Variable names beginning with one of the other letters (*a* through *h,* and *o* through *z*), are implicitly used to store real values. Thus, with implicit typing, amount represents a real value and money represents an integer value. An easy way to remember which letters are used for integers is to observe that the range of letters is *i* through *n,* which are the first two letters of the word *integer.*

With *explicit typing,* Fortran statements are used to specify the variable types. For example, the statements

```
integer :: width
real :: item, length
```

specify that width is a variable containing an integer value and that item and length are variables containing real values. These specification statements or *type statements* have the following general forms:

```
integer :: variable list
```

```
real :: variable list
```

The variable list contains variable names separated by commas. These are *nonexecutable statements* because they are not translated into machine language. Instead, they are used by the compiler to assign memory locations and to specify the types of values to be stored in the locations. Variable names can be listed in any order; in our programs we will use alphabetical order.

Initial values can also be assigned to variables in the type statement. For example, the following statement assigns memory locations for integers count and degrees, and also *initializes* the variable count to a value of zero and initializes the value of degrees to 180:

```
integer :: count=0, degrees=180
```

Do not assume that a variable is initialized to a value such as zero when a program is executed unless a statement in the program initializes the variable.

You will find it helpful to select a variable name that is descriptive of the value being stored. For example, if a value represents a tax rate, name it rate or tax_rate. If the implicit typing of the variable name does not match the type of value to be stored in it, then use a real or integer statement at the beginning of your program to specify the desired type of value. In fact, it is good programming style to list all variables in specification statements, including those correctly typed by the implicit rules. Fortran programs can also include the following specification statement that informs the compiler to require explicit typing for all variables:

```
implicit none
```

This statement is used in the example programs in this module, and thus all variables are included in specification statements.

The parameter statement is a specification statement used to assign names to constants; it has the following general form:

```
parameter (name1 = expression, name2 = expression, . . . )
```

The expression after the equal sign typically is a constant, although it can be an expression consisting of other constants and operations such as those discussed later in this chapter. An example of a parameter statement is

```
real :: pi
parameter (pi=3.141593)
```

Type statements should precede the parameter statement in order to assign the proper type to the constant name. The value of a constant defined with the parameter statement cannot be changed within a program.

Scientific Notation

When a real number is very large or very small, decimal notation does not work satisfactorily. For example, a value that is used frequently in chemistry is Avogadro's constant, whose value to four significant places is 602,300,000,000,000,000,000,000. Obviously, we need a more manageable notation for very large values like Avogadro's constant or for very small values like 0.000000000042. *Scientific notation,* commonly used in science, expresses a value as a number between 1 and 10 multiplied by a power of 10. In scientific notation Avogadro's constant becomes 6.023×10^{23}. Elements of this form are commonly referred to as the mantissa (6.023) and the exponent (23). The Fortran form of scientific notation, called *exponential notation,* expresses a value as a number between 0.1 and 1 multiplied by an appropriate power of 10. Exponential notation uses the letter E to separate the mantissa and the exponent. In exponential form Avogadro's constant becomes 0.6023E24. The following are other examples of decimal values in scientific and exponential notation:

Decimal	Scientific	Exponential
3,876,000,000	3.876×10^9	0.3876E10
0.0000010053	1.0053×10^{-6}	0.10053E$-$05
$-8,030,000$	-8.03×10^6	-0.803E07
-0.000157	-1.57×10^{-4}	-0.157E-03

Although Fortran uses an exponential form with a mantissa between 0.1 and 1.0, it does accept mantissas outside that range; for instance, the constant 0.16E03 is also valid in the forms 1.6E02 or 0.016E04.

Magnitude Limitations

The magnitude and precision of values that can be stored in a computer are limited. All limitations on values depend on the specific computer. For instance, π is an irrational number and cannot be written exactly with a finite number of decimal positions; in a computer with seven digits of accuracy, π can be stored as 3.141593. In addition to limits on the number of significant positions in the mantissa, there are also limits on the size of the exponent.

Table 2-1 compares the approximate range of integers that can be stored in several computers. The range of values is determined by the design of the

Table 2-1   Integer Representations in Typical Computers

Computer	Number of Binary Digits (or Bits) Per Data Value	Number of Integers*
Texas Instruments 32020 Microprocessor	16	65,536
Motorola 68020 Microprocessor	32	4.3×10^9
IBM PC	32	4.3×10^9
Macintosh	32	4.3×10^9
VAX 11/780	32	4.3×10^9
Sun Sparc Station 10	32	4.3×10^9
Cray Y MP C90	48	2.8×10^{14}
Cray-2	64	1.8×10^{19}

* Generally, half of the integers will represent negative values, so the maximum integer is usually equal to the number of integers divided by 2.

central processing unit (CPU). Check a reference manual to find the ranges of real and integer values for the computer you are using.

Try It

Try this self-test to check your memory of some key points from Section 2-1. If you have any problems with the exercises, you should reread this section. The solutions are given at the end of this module.

Problems 1–10 contain both valid and invalid variable names. Explain why the invalid names are unacceptable.

1. sq.yd	2. micron
3. WEIGHT	4. Degrees
5. NetWt	6. side__1
7. f(t)	8. 3J6
9. total	10. five%

In problems 11–16 tell whether or not the pair of real constants represents the same number. If not, explain.

11. 15.7; 0.157E2	12. −1.7; 1.7E−01
13. 10.0; 1000.0E−02	14. 0.005; 0.00005E02
15. 0.899; 89.9E02	16. −0.044; 4.4E−02

2-2   ARITHMETIC OPERATIONS

A computation in Fortran may be specified with an *assignment statement*, which has the general form

$$\boxed{variable\ name = expression}$$

The simplest form of an expression is a constant. The following assignment statement assigns a value of 5.5 to the variable voltage:

```
real :: voltage
...
voltage = 5.5
```

The name of the variable receiving a new value must always be on the left side of the equal sign. In Fortran the equal sign can be read as "is assigned the value of." Thus, this statement could be read "voltage is assigned the value 5.5." The term *initialized* is often used to describe the first value assigned to a variable in a program; thus, this statement could be read "voltage is initialized to the value 5.5." This assignment could also have been made in the specification statement

```
real :: voltage=5.5
```

Table 2-2   Arithmetic Operations in Algebraic Form and in Fortran

Operation	Algebraic Form	Fortran
Addition	$a + b$	a + b
Subtraction	$a - b$	a - b
Multiplication	$a \times b$	a*b
Division	$\dfrac{a}{b}$	a/b
Exponentiation	a^3	a**3

Simple Arithmetic

Often we want to calculate a new value using arithmetic operations with other variables and constants. For instance, assume that the variable radius has been assigned a value, and we want to calculate the area of a circle having that radius. To do so, we must square the radius and then multiply by the value of π. Table 2-2 shows the Fortran expressions for the basic arithmetic operations. Note that an asterisk (instead of ×) represents multiplication; this avoids confusion, because a×b (commonly used in algebra to indicate the product of a and b) represents a variable name in Fortran. Division and exponentiation also have different symbols that allow us to write these arithmetic operations on a single line. The operations shown in Table 2-2 are called binary operations because they are operations between two values.

Evaluation of Arithmetic Expressions

Because several operations can be combined in one arithmetic expression, it is important to determine the precedence of the operations, that is, the order in which the operations are performed. For instance, consider the following assignment statement that calculates the area of a circle:

```
real :: area, pi=3.14159, radius
...
area = pi*radius**2
```

If the exponentiation is performed first, we compute pi($radius^2$); if multiplication is performed first, we compute (pi·radius)^2. Note that the two computations yield different results. The precedence of computations in Fortran is given in Table 2-3 and follows standard algebraic priorities. From this table we can determine the order of the operations in the previous statement; radius is first squared, and then the result is multiplied by pi — correctly determining the area of the circle. (We assume that both pi and radius have been initialized by previous statements in the program.)

If a minus sign or a plus sign is associated with a single value, it is a unary operator because it operates on a single value. For example, in the expression –a + b, the minus sign is a unary operator and the plus sign is a binary operator. From Table 2.3 we see that unary operations are performed before the binary operations of addition and subtraction and after all other opera-

Table 2-3 Precedence of Arithmetic Operators

Precedence	Operator	Associativity
1	Parentheses ()	Innermost first
2	Exponentiation	Right to left
3	Binary operators * /	Left to right
4	Unary operators + −	Left to right
5	Binary operators + −	Left to right

tions. For example, $-a**2$ is computed as if it were $-(a**2)$, $-a*b$ is computed as if it were $-(a*b)$, and $-a+b$ is computed as if it were $(-a) + b$.

When two operations are on the same priority level, as in addition and subtraction, all operations except exponentiation are performed from left to right. Thus, b–c+d is evaluated as $(b-c) + d$. If two exponentiations occur sequentially in Fortran, as in a**b**c, they are evaluated right to left, as in a**(b**c). Thus, 2**3**2 is 2^9, or 512, as opposed to (2**3)**2, which is 8^2, or 64.

A more complicated example is represented by the following equation for one of the real roots of a quadratic equation:

$$x_1 = \frac{-b + \sqrt{b^2 - 4ac}}{2a}$$

where a, b, and c are coefficients of a quadratic equation $(ax^2 + bx + c = 0)$. Because computers cannot divide by zero, we will assume for now that a is not equal to zero. The value of x_1 can be computed in Fortran with the following statement, assuming that the real variables a, b, and c have been initialized:

```
real :: a, b, c
...
x1 = (-b + (b**2 - 4.0*a*c)**0.5)/(2.0*a)
```

To check the order of operations in a long expression, we should start with the operations inside parentheses; that is, find the operation done first, then second, and so on. The following diagram outlines this procedure, using braces to show the steps of operations. Beneath each brace is the value calculated in that step:

$$x1 = \underbrace{(-b}_{-b} + \underbrace{\overbrace{(b**2 - 4.0*a*c)}^{b^2 - 4ac}**0.5)}_{-b + \sqrt{b^2 - 4ac}}/\underbrace{(2.0*a)}_{2a}$$

$$\frac{-b + \sqrt{b^2 - 4ac}}{2a}$$

As shown in the final brace, the desired value is computed by this expression.

Parentheses placement is important. If the outside set of parentheses in the numerator in the previous Fortran statement were omitted, the assignment statement would become

$$x1 = \underbrace{-b}_{-b} + \underbrace{\overbrace{(b**2 \underbrace{- 4.0*a*c}_{b^2 - 4ac})**0.5}^{\sqrt{b^2 - 4ac}}/\underbrace{(2.0*a)}_{2a}}_{\dfrac{\sqrt{b^2 - 4ac}}{2a}}$$

$$-b + \dfrac{\sqrt{b^2 - 4ac}}{2a}$$

As you can see, omission of the outside set of parentheses causes the wrong value to be calculated as a root of the original quadratic equation. Omitting necessary parentheses results in incorrect calculations. Using extra parentheses to emphasize the order of calculations is recommended, even though they may not be required. We often insert extra parentheses in a statement to make the statement more readable.

You also may want to break a long statement into several smaller statements. The expression b^2-4ac in the quadratic equation is called the discriminant. Both roots of the solution could be calculated with the following statements after initialization of a, b, and c:

```
real :: a, b, c, discriminant, x1, x2
...
discriminant = b**2 - 4.0*a*c
x1 = (-b + discriminant**0.5)/(2.0*a)
x2 = (-b - discriminant**0.5)/(2.0*a)
```

In the preceding statements we assume that discriminant is positive, enabling us to obtain x1 and x2, the two real roots to the equation. If the discriminant is negative, an execution error occurs when we attempt to compute the square root of the negative value. If the value of a is zero, we get an execution error for attempting to divide by zero. Later chapters present techniques for handling these situations.

Truncation and Mixed-Mode Operations

When an arithmetic operation is performed using two real numbers, its *intermediate result* is a real value. For example, we can calculate the circumference of a circle using the following statements:

```
real :: circumference, diameter, pi=3.141593
...
circumference = pi*diameter
```

In both statements we have multiplied two real values, giving a real result, which is then stored in the real variable circumference.

Similarly, arithmetic operations between two integers yield an integer. For instance, if i and j represent two integers, and if i is less than or equal to j, then we can calculate the number of integers included in the closed interval [i,j] with the following statement:

```
integer :: i, interval, j
...
interval = j - i + 1
```

Thus, if i = 5 and j = 11, interval will be assigned the value 7, which is the number of integers in the interval [5,11].

Now consider these statements:

```
integer :: length
real :: side
...
length = side*3.5
```

We know that the multiplication between the real value side and the real constant 3.5 yields a real result. In this case, however, the real result is stored in an integer variable. For example, if side contains the value 1.1, then length will be computed as 3, not the correct value 3.85. When the computer stores a real number in an integer variable, it ignores the fractional portion and stores only the whole number portion of the real number; this process is called *truncation*.

Computations with integers can also give unexpected results. Consider the following statement, which computes the average, or mean, of two integers, n1 and n2:

```
integer :: mean, n1, n2
...
mean = (n1 + n2)/2
```

If we assume that all the variables in the statement are integers, the result of the expression will be an integer. Thus, if n1 = 2 and n2 = 4, the mean value will be the expected value, 3. But if n1 = 2 and n2 = 3, the result of the division of 5 by 2 will be 2 instead of 2.5 because the division involved two integers; hence, the intermediate result must be an integer. At first glance it might seem that we could solve this problem if we stored the average in a real variable named average and used this statement:

```
integer :: n1, n2
real :: average
...
average = (n1 + n2)/2
```

Unfortunately, this cannot correct our answer. The result of integer arithmetic is still an integer; all we have done is move the integer result into a real variable. Thus, if n1 = 2 and n2 = 3, then (n1 + n2)/2 = 2 and the value stored in average will be 2.0, not 2.5. One way to correct this problem is to declare n1 and n2 to be real values and use the following statement to calculate the average:

```
real :: average, n1, n2
...
average = (n1 + n2)/2.0
```

Note the difference between *rounding* and truncation. With rounding, the result is the integer closest in value to the real number. Truncation, however, causes any decimal portion to be dropped. If we divide the integer 15 by the integer 8, the truncated result is 1, and the rounded result is 2.

The effects of truncation can also be seen in the following statement, which appears to calculate the square root of number:

```
real :: number, root
...
root = number**(1/2)
```

However, since 1/2 is truncated to 0, we are really raising number to the zero power; root will always contain the value 1.0, no matter what value is in number.

We have seen that an operation involving only real values yields a real result, and an operation involving only integer values yields an integer result. Fortran also accepts a *mixed-mode operation,* which is an operation involving an integer value and a real value. The intermediate result is a real value. The final result depends on the type of variable used to store the result of the mixed-mode operation. Consider the following assignment statement for computing the perimeter of a square whose side is a real value:

```
real :: perimeter, side
...
perimeter = 4*side
```

This multiplication is a mixed-mode operation between the integer constant 4 and the real variable side. The intermediate result is real and is correctly stored in the real variable perimeter.

Using a mixed-mode operation, we can now correctly calculate the square root of the integer number by using this statement:

```
integer :: number
real :: root
...
root = number**0.5
```

The mixed-mode exponentiation yields a real result, which is stored in a real variable root.

To compute the area of a square we could use either of the following statements assuming that side and area are real variables:

```
area = side**2
area = side**2.0
```

The result in both cases is real, but the first form is preferable because exponentiation to an integer power is generally performed internally in the computer with a series of multiplications such as side times side. If an exponent is real, however, the operation is performed by the arithmetic logic unit (ALU) using logarithms; side**2.0 is actually computed as antilog

$(2.0 \times \log(\texttt{side}))$. Logarithms can introduce small errors into calculations; although `5.0**2` is always 25.0, `5.0**2.0` is often computed as 24.99999. Also, note that `(-2.0)**2` is a valid operation, but `(-2.0)**2.0` is an invalid operation — the logarithm of a negative value does not exist, and an execution error occurs. As a general rule, when raising numbers to an integer power, you should use an integer exponent, even though the base number is real.

Assume that we want to calculate the volume of a sphere. The volume is computed by multiplying 4/3 times π times the radius cubed. Thus, if the value of the radius is contained in the variable `radius`, the following mixed-mode statement at first appears correct:

```
real :: radius, volume
...
volume = (4/3)*3.141593*radius**3
```

The expression contains integer and real values, so the result will be a real value. However, the division of 4 by 3 yields the intermediate value of 1, not 1.333333; therefore, the final answer will be incorrect.

Because mixed-mode operations can sometimes give unexpected results, you should avoid writing arithmetic expressions that include them. A mixed-mode expression is desirable only in an exponentiation operation in which a real value is raised to an integer power.

Underflow and Overflow

In a previous section we discussed magnitude limitations for the values stored in variables. Because the maximum and minimum values that can be stored in a variable depend on the computer system itself, a computation may yield a result that can be stored in one computer system but is too large to be stored in another. For example, suppose we execute the following assignment statements:

```
real :: x, y, z
...
x = 0.25E20
y = 0.10E30
```

These values are both valid for a computer with an exponent range of -38 through 38. Suppose we now execute the following statement:

```
z = x*y
```

The numerical result of this multiplication is `0.025E50`. Clearly this result is too large to store in a computer with a maximum exponent of 38. The error is not a syntax error; the statements themselves are valid Fortran 90 statements. The error is a logic error because it occurred during execution of the program, and it is called an exponent *overflow* error because the exponent of the result of an arithmetic operation is too large to store in the computer's memory.

Exponent *underflow* is a similar error caused by the exponent of the result of an arithmetic operation being too small to store in the computer's memory. Using the same computer as we did for the previous example, the following statements would generate an exponent underflow error because the value that should be computed for c (2.50E–40) has an exponent smaller than − 38:

```
real :: a, b, c
...
a = 0.25E-20
b = 0.10E+20
c = a/b
```

If you get exponent underflow or overflow errors when you run your programs, you need to examine the magnitude of the values you are using. If you really need values whose exponents exceed the limits of your computer, the only solution is to switch to a computer or variable type that can handle a wider range of exponents. Most of the time, exponent underflow and overflow errors are caused by errors other than the computer capacity in a program. For example, if variables are initialized to incorrect values or the wrong arithmetic operation is specified, exponent underflow or overflow can occur, but the source of the problem is elsewhere in the program.

Try It

Try this self-test to check your memory of some key points from Section 2-2. If you have any problems with the exercises, you should reread this section. The solutions are given at the end of this module.

1. What value is stored in y after the following statements are executed?

```
real :: a=2.0, x=5.0, y
...
y = x + 3.0/a*5.0
```

2. What value is stored in x after the following statements are executed?

```
real :: x, x1=4.0, x2=10.0
...
x = 1.0/(1.0/x1) + (1.0/x2)
```

3. What value is stored in result after the following statements are executed?

```
integer :: b=16, result, x=5, y=-21
...
result = x + y + b/5
```

2-3   INTRINSIC FUNCTIONS

Algorithms frequently require common engineering functions, such as the square root of a value, the absolute value of a number, or the sine of an angle. Because these operations occur so frequently, built-in computer functions called *intrinsic functions* are available to handle these routine computations. Instead of using the arithmetic expression x**0.5 to compute a square root, we can use the intrinsic function sqrt(x). Similarly, we can refer to the absolute value of b as abs(b). A list of some commonly used

Table 2-4   Common Intrinsic Functions

Function Name and Argument	Function Value	Comments
sqrt(x)	$\sqrt{x}$	Square root of x
abs(x)	$\lvert x \rvert$	Absolute value of x
sin(x)	sine of angle x	x must be in radians
cos(x)	cosine of angle x	x must be in radians
tan(x)	tangent of angle x	x must be in radians
exp(x)	e^x	e raised to the x power
log(x)	$\log_e x$	Natural log of x
log10(x)	$\log_{10} x$	Common log of x
int(x)	Integer part of x	Converts a real value to an integer value
real(i)	real value of i	Converts an integer value to a real value
mod(i,j)	integer remainder of i/j	Remainder or modulo function

intrinsic functions appears in Table 2-4. Appendix A contains an expanded list of common Fortran 90 intrinsic functions with brief descriptions.

The name of a function is followed by the input to the function, also called the *argument* of the function, which is enclosed in parentheses. This argument can be a constant, a variable, or an expression. For example, suppose we want to compute the cosine of the variable `angle` and store the result in another variable `cosine`. From Table 2-4 we see that the cosine function assumes that its argument is in radians. If the value in `angle` is in degrees, we must change the degrees to radians (1 degree = $\pi/180$ radians) and then compute the cosine. The following statement performs the degree-to-radian conversion within the function argument. Note that the inside set of parentheses is not required but emphasizes the conversion factor.

```
real :: angle, cosine
...
cosine = cos(angle*(3.141593/180.0))
```

The `real` and `int` functions should be used to avoid errors and undesirable mixed-mode arithmetic expressions by explicitly converting variable types. For example, if we are computing the average of a group of real values, we need to divide the real sum of the values by the number of values. We can convert the integer number of values into a real value for the division using the `real` function, as shown in the following statement:

```
integer :: n
real :: average, sum
...
average = sum/real(n)
```

If the values were integers, the sum would also be an integer. We probably would still want the average to be represented by a real value, which could be specified by

```
integer :: n, sum
real :: average
...
average = real(sum)/real(n)
```

Many intrinsic functions are *generic functions*, which means that the value returned is the same type as the input argument. The absolute value function abs is a generic function. If x is an integer, then abs(x) is also an integer; if x is a real value, then abs(x) is also a real value. Some functions specify the type of input and output required. For example, iabs is a function that requires an integer input and returns an integer absolute value from the function. If k is an integer, then abs(k) and iabs(k) return the same value. Appendix A contains all forms of the numerical intrinsic functions and identifies all generic functions.

Try It

Try this self-test to check your memory of some key points from Sections 2-2 and 2-3. If you have any problems with the exercises, you should reread these sections. The solutions are given at the end of this module.

In problems 1–4 convert the equations into Fortran assignment statements. Assume all variables represent real values.

1. Magnitude:

$$\text{magnitude} = \sqrt{x^2 + y^2}$$

2. Velocity addition formula:

$$u = \frac{u + v}{1 + \dfrac{u \cdot v}{c^2}}$$

3. Damped harmonic motion:

$$y = y_0 \cdot e^{-at} \cdot \cos(2\pi f \cdot t)$$

4. Temperature conversion to degrees K:

$$T = \left(\frac{5}{9}(T_f - 32) \right) + 273.15$$

In problems 5–10, convert the Fortran statements into algebraic form.

5. Potential energy:

```
potential_energy = -g*me*m/r
```

6. Electric flux:

```
flux = e*da*cos(theta)
```

7. Average velocity:

```
average_velocity = (x2 - x1)/(t2 - t1)
```

8. Centripetal acceleration:

```
pi = 3.141593
centripetal_acc = 4.0*pi**2*r/(t**2)
```

9. Distance of an accelerating body:

```
distance = velocity*time + acceleration*time**2/2.0
```

10. Atmospheric pressure adjusted for elevation:

```
pressure = p0*exp(-m*g*x/r*tk)
```

2-4 SIMPLE INPUT AND OUTPUT

Before discussing input and output statements in this section, we outline the proper form for entering Fortran statements in a program. We then present several types of I/O statements. A computer can accept input from different sources, but in this chapter we assume that the input is from the keyboard. Similarly, we can direct output to different devices, but in this chapter we assume that the output is directed to a terminal screen or a printer. Input and output using a data file are discussed in Chapter 4.

Fortran Source Forms

We can create Fortran programs using an editor or a word processor to enter the text. Each statement is generally entered on a new line. Most Fortran statements are short, but occasionally a statement will be long enough to require more than one line. *Statement numbers,* which are nonzero positive integers that precede the Fortran statement, are required by only a few statements. Fortran also allows comment lines in a program; *comments* are used for documenting the steps in a program, but the comments are not converted into machine language or used during execution of the program. It is good programming practice to include several comment lines near the beginning of the program to describe its purpose. In fact, you might want to include comments containing the information in step 1 (problem statement) and step 2 (input/output description) of the problem-solving process. Blank lines can also be included anywhere in the program.

We refer to specific positions within a line by column number: The first position in a line is column 1, the second position in a line is column 2, and so on. In general, blanks can be inserted anywhere in the statement for readability. Fortran 90 allows two types of source forms, free-source form and fixed-source form. A program must be in one or the other form; it cannot mix the forms in the same program. We now discuss the details of both forms so that you are familiar with the differences; we will use free-source form in the example programs in this module.

In *fixed-source form,* a line can contain no more than 72 characters. Positions 1 through 5 are reserved for Fortran statement labels. Any nonblank character except a zero in position 6 indicates that a line is a *continuation* of the previous line. Fortran statements must be within positions 7 through 72. Comment lines are indicated by the letter C or the character * in position 1. Comments can also be indicated by the character !; a comment initiated by the character ! can be on a line by itself or can follow a Fortran statement.

In *free-source form,* a line can contain no more than 132 characters, and there are no restrictions on where a statement may appear in the line. The character & is used at the end of the current line to indicate that the current statement is to be continued on the next line. Comments are indicated by the character ! and can appear on lines by themselves or can follow a Fortran statement.

In both fixed-source and free-source forms, a semicolon can be used to separate multiple statements on the same line, as in

```
count = 0; sum = 0
```

A trailing semicolon is ignored.

List-Directed Output

Fortran has two types of statements that allow us to perform I/O operations. *List-directed I/O* statements are easy to use but give us little control over the exact spacing used in the input and output lines. *Formatted I/O*, although more involved, allows us to control the input and output forms with greater detail. This section presents list-directed input/output and simple formatted output. For the example programs in this module, we will use list-directed input statements and formatted output statements. This technique allows us to be flexible about the form we use for information we read and, at the same time, allows us to be specific about the form used to display the information computed by our program.

If we wish to print the value stored in a variable, we must tell the computer the variable's name. The computer can then access the memory location and print its contents. The general form of the list-directed print statement is

```
print*, expression list
```

The expressions in the list must be separated by commas. The corresponding values are printed in the order in which they are listed in the print statement. The output from each print statement begins on a new line. The number of decimal positions printed for real values and the spacing between items vary depending on the compiler used. If a value to be printed is very large or very small, many compilers automatically print the value in exponential notation instead of in decimal form. In our examples we assume that values are separated by spaces and that six decimal places are printed for real values.

In the examples that follow, computer output is shown inside a rounded box, as illustrated in Example 2-1.

EXAMPLE 2-1 ## Weight, Volume, and Density

Print the stored values of the variables weight, volume, and density. Assume the following type statements:

```
integer :: weight
real :: density, volume
```

SOLUTION

Computer Memory

weight	35000
volume	3.15
density	0.0000156

Fortran Statement

```
print*, weight, volume, density
```

Computer Output

```
35000 3.150000 0.156000E-04
```

. . . .

Descriptive information (sometimes called literal information or *literals*) may also be included in the expression list by enclosing the information in quotation marks or in apostrophes. An apostrophe in a literal is represented by two apostrophes, as in "User''s program", and quotation marks are represented by two quotation marks. An uppercase character is distinguished from a lowercase character in a literal; thus, the following literals are all different: "time", "Time", and "TIME". The descriptive information is printed in the output line along with the values of any variables; we also assume that spaces are inserted after literals in the output line.

EXAMPLE 2-2 | **Literal and Variable Information**

Print the value of the real variable rate with a literal that identifies the value as a flow rate in gallons per second.

SOLUTION

Computer Memory

rate | 0.065

Fortran Statement

```
print*, "flow rate is", rate, "gallons per second"
```

Computer Output

```
flow rate is 0.065000 gallons per second
```

. . . .

List-Directed Input

We frequently need to read information with our programs. The general form of a list-directed read statement is

```
read*, variable list
```

The variable names in the list must be separated by commas. The variables receive new values in the order in which they are listed in the read statement. These values should agree in type (integer or real) with the variables in the list. The system will wait for you to enter the appropriate data values when the read statement is executed. If more than one value is being read by the statement, the data values can be separated by commas or blanks, or they can be entered on separate lines.

A read statement will read as many lines as needed to determine values for the variables in its list. Therefore, if a read statement has four variables in it, the computer will wait until you have entered four values; the values could be on one line (separated by commas or blanks) or on several lines. Also, each read statement begins reading from a new line. Thus, if you have executed a read statement with four variables in its list and you entered five values on one line, the last value will not be read; the next read statement will assume that a new line is used for entering information.

EXAMPLE 2-3

Capacitor Charge

Read the beginning and ending charges for a capacitor. Assume the following type statement:

```
real :: final_charge, initial_charge
```

SOLUTION 1

This solution uses one read statement; thus, both values can be entered on the same line.

Fortran Statement

```
read*, initial_charge, final_charge
```

Data Line

```
186.93, 386.21
```

Computer Memory

initial_charge	**186.93**
final_charge	**386.21**

The data values in this example could also have been entered on separate lines since the read statement will read as many lines as needed to find values for the variables in its list.

SOLUTION 2

This solution uses two read statements; thus, the values must be entered on different lines.

Fortran Statements

```
read*, initial_charge
read*, final_charge
```

Data Lines

```
186.93
386.21
```

Computer Memory

initial__charge   `186.93`

final__charge   `386.21`

. . . .

Formatted Output

To specify the form in which data values are printed and where on the output line they are printed requires formatted statements. The general form of a formatted `print` statement is

> `print k,` *expression list*

The expression list designates the memory locations whose contents will be printed or arithmetic expressions whose values will be printed. The list of expressions also determines the order in which the values will be printed. The reference k refers to a `format` statement that will specify the spacing to be used in printing the information. A sample `print` and `format` combination is

```
      real :: distance, time
      ...
      print 5, time, distance
    5 format(1x,f5.1,2x,f7.2)
```

The general form of the `format` statement is

> `k format` *(specification list)*

The specifications in the list tell the computer both the vertical and horizontal spacing to be used when printing the output information. The vertical spacing options include printing on the top of a new page (if the output is being printed on paper), printing on the next line (single spacing), double spacing, and no spacing. Horizontal spacing includes indicating how many digits will be used for each value, how many blanks will be between numbers, and how many values are to be printed per line.

To understand the specifications used to describe the vertical and horizontal spacing, we must first examine the output of a printer or terminal — the most common output devices. Other forms of output have similar characteristics.

The line printer prints on continuous computer paper that is perforated so it is easy to separate the pages; a laser printer uses individual sheets of paper. Typically, 55 to 75 lines of information can be printed per page. The number of characters printed per line depends on the font size, but a typical line contains 80 to 132 characters. The `print/format` combination specifically describes where each line is to be printed on the page (vertical spacing) and which positions in the line will contain data (horizontal spacing).

The computer uses the specification list to construct each output line internally in memory before actually printing the line. This internal memory region is called a *buffer.* The buffer is automatically filled with blanks before it is used to construct a line of output. The first character of the buffer is called the *carriage control character;* it determines the vertical spacing for the line. The remaining characters represent the line to be printed, as shown in the following diagram:

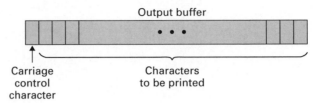

The following list shows some of the valid carriage control characters and the vertical spacing they generate:

Carriage Control Character	Vertical Spacing
1	New page
blank	Single spacing
0	Double spacing

Double spacing causes one line to be skipped before the current line of output is printed. On most computers an invalid carriage control character causes single spacing.

A terminal screen does not have the same capabilities as a printer for spacing; therefore, carriage control is usually ineffective and causes single spacing. If the terminal I/O does not use carriage control, then the entire buffer, including the carriage control character, will appear on the terminal screen.

We now examine five format specifications that describe how to fill the output buffer. Commas are used to separate specifications in the format statement. When needed for clarity in either format statements or buffer contents, a blank is indicated by the character b placed one-half space below the regular line. Additional format specifications are included in Section 2-8.

Literal Specification

The literal specification allows us to put characters directly into the buffer. The characters must be enclosed in double quotation marks or in single quotation marks. These characters can represent the carriage control character or the characters in a literal. The following examples illustrate use of the literal specification in format statements.

EXAMPLE 2-4

Title Heading

Print the title heading Test Results on the top of a new page, *left-justified* (that is, no blanks to the left of the heading).

SOLUTION

Fortran Statements

```
print 4
4 format("1","Test Results")
```

Buffer Contents

```
1Test Results
```

Computer Output

```
        111
123456789012
```

```
Test Results
```

The row of small numbers above the computer output shows the specific column of the output line: T is in column 1, e is in column 2, and so on. The buffer is filled according to the format statement. No variable names are listed on the print statement; hence, no values are printed. The literal specifications cause the characters 1Test Results to be put in the buffer, beginning with the first position in the buffer. After filling the buffer, as instructed by the format statement, the carriage control determines vertical spacing but is not printed. The character 1 in the carriage control position tells the computer to begin a new page. The rest of the buffer is then printed.

. . . .

| EXAMPLE 2-5 |

Column Headings

Double-space from the last line printed, and print column headings 1988 kW-hrs and 1989 kW-hrs, with no blanks on the left side of the line and seven blanks between the two column headings.

CORRECT SOLUTION

Fortran Statements

```
print 3
3 format ("0","1988 kWhbbbbbbb1989 kWh")
```

Buffer Contents

```
01988 kWhbbbbbbb1989 kWh
```

Computer Output

```
            11111111112222
1234567890123456789 0123
```

```
1988 kWh      1989 kWh
```

The line shown is printed after double spacing from the previous line of output.

INCORRECT SOLUTION

Fortran Statements

```
        print 3
 3 format("1988 kWhbbbbbbb1989 kWh")
```

Buffer Contents

```
1988 kWhbbbbbbb1989 kWh
```

Computer Output

```
         11111111112222
12345678901234567890123
```

```
988 kWh        1989 kWh
```

In this example we forgot to specify the carriage control. However, the computer does not forget: The first position of the buffer contains a 1, which indicates spacing to a new page. The rest of the buffer is then printed.

· ·

x Specification

The x specification inserts blanks into the buffer. Its general form is nx, where n represents the number of blanks to be inserted in the buffer. An example using both the x specification and the literal specification follows.

EXAMPLE 2-6

Centered Heading

Print the heading Experiment 1 centered at the top of a new page.

SOLUTION

Assume that an output line contains 65 characters. To center the heading, we determine the number of characters in the heading (12), subtract that from 65 (65 − 12 = 53), and divide that by 2 to put one-half the blanks in front of the heading (53/2 = 26).

Fortran Statements

```
        print 35
 35 format("1",26x,"Experiment 1")
```

Buffer Contents

```
1bbbbbbbbbbbbbbbbbbbbbbbbbbExperiment 1
```

Computer Output

```
222333333333
789012345678
```

Experiment 1

. . . .

i Specification

The literal specification and the x specification allow us to specify carriage control and to print headings. They cannot, however, be used to print variable values. We now examine a specification that prints the contents of integer variables: The specification form is iw, where w represents the number of positions (width) to be assigned in the buffer for printing the value of an integer variable. The value is always *right-justified* (no blanks to the right of the value) in those positions in the buffer. Extra positions on the left are normally filled with blanks. Thus, if the value 16 is printed with an i4 specification, the four positions contain two blanks followed by 16. If not enough positions are available to print the value, including a minus sign if the value is negative, the positions are filled with asterisks. Hence, if we print the value 132 or -14 with an i2 specification, the two positions are filled with asterisks. It is important to recognize that the asterisks do not necessarily indicate that there is an error in the value; the asterisks may indicate that you need to assign a larger width in the corresponding output specification.

More than one variable name is often listed in the print statement. When interpreting a print/format combination, the compiler will match the first variable name to the first specification for printing values, and so on. Therefore, there should be the same number of specifications for printing values as there are variables on the print statement list. (In Section 2-8 we explain what happens if the number of data specifications does not match the number of variables.)

EXAMPLE 2-7   ## Integer Values

Print the values of the integer variables sum, mean, and n on the same line, single-spaced from the previous line.

SOLUTION

Computer Memory

sum	**12**
mean	**−14**
n	**−146**

Fortran Statements

```
        print 30, sum, mean, n
   30 format (1x,i3,2x,i2,2x,i4)
```

Buffer Contents

```
bb12bb**bb−146
```

Computer Output

```
          1111
1234567890123
```

```
12   **   −146
```

The computer will print sum with an i3 specification, mean with an i2 specification, and n with an i4 specification. The value of sum is 12, so the three
corresponding positions contain a blank followed by the number 12. The
value of mean, −14, requires at least three positions, so the two specified
positions are filled with asterisks. The value of n fills all four allotted positions. The carriage control character is a blank, thus the line of output is
single-spaced from the previous line.

. . . .

EXAMPLE 2-8

Literal and Variable Information

On separate lines print the values of mean and sum, along with an indication
of the name of each of the integer variables.

SOLUTION

Computer Memory

sum   **12**

mean **−14**

Fortran Statements

```
   print 2, mean
 2 format(1x,"mean = ",i4)
   print 3, sum
 3 format(1x,"sum = ",i4)
```

Buffer Contents

```
bmeanb=bb−14
```

```
bsumb=bbb12
```

Computer Output

```
         11
12345678901
```

```
mean =   −14
sum =    12
```

. . . .

f Specification

The f specification is used to print real numbers in a decimal form (for example, 36.21) as opposed to an exponential form (for example, 0.3621E+02). The general form for an f specification is fw.d, where w represents the total width (number of positions, including the decimal point) to be used, and d represents the number of those positions that will represent decimal positions to the right of the decimal point, as shown:

$$\text{decimal portion} = d$$
$$\overbrace{\texttt{xx.xxx}}$$
$$\text{total width} = w$$

If the value to be printed has fewer than d decimal positions, zeros are inserted on the right side of the decimal point. Thus, if the value 21.6 is printed with an f6.3 specification, the output is 21.600 If the value to be printed has more than d decimal positions, only d decimal positions are printed, dropping the rest. Thus, if the value 21.86342 is printed with an f6.3 specification, the output is 21.863. Most compilers will round to the last decimal position printed; in these cases the value 18.98662 is printed as 18.987 if an f6.3 specification is used.

If the integer portion of a real value requires fewer positions than allotted in the f specification, the extra positions on the left side of the decimal point are filled with blanks. Thus, if the value 3.123 is printed with an f6.3 specification, the output is a blank followed by 3.123. If the integer portion of a real value, including the minus sign if the value is negative, requires more positions than allotted in the f specification, the entire field is filled with asterisks. Thus, if the value 312.6 is printed with an f6.3 specification, the output is ******.

If a value is between −1 and +1, positions must usually be allowed for both a leading zero to the left of the decimal point and a minus sign if the value is negative. Thus, the smallest f specification that could be used to print −0.127 is f6.3. If a smaller specification width were used, all the positions would be filled with asterisks.

EXAMPLE 2-9

Angle theta

Print the value of a real variable called theta. The output should be in this form:

$$\text{theta} = \text{XX.XX}$$

SOLUTION

Computer Memory

theta **3.182**

Fortran Statements

```
print 1, theta
1 format(1x,"theta = ",f5.2)
```

Buffer Contents

> bthetab=bb3.18

Computer Output

```
          1111
1234567890123
```

> theta = 3.18

. . . .

e Specification

Real numbers may be printed in an exponential form with the e specification. This specification is used primarily for very small or very large values, or when you are uncertain of the magnitude of a number. If you use an f format that is too small for a value, the output field will be filled with asterisks. In contrast, a real number will always fit in an e specification field.

The general format for an e specification is ew.d. The w again represents the total width or number of positions to be used in printing the value. The d represents the number of positions to the right of the decimal point, assuming that the value is in exponential form, with the decimal point to the left of the first nonzero digit. The framework for printing a real value in an exponential specification with three decimal places is

$$
\begin{array}{c}
\text{decimal portion} = d \\
\overbrace{} \\
\text{s0.xxxEsxx} \\
\underbrace{} \\
\text{total width} = w
\end{array}
$$

The symbol s indicates that positions must be reserved for both the sign of the value and the sign of the exponent in case they are negative. Note that, with all the extra positions, the total width becomes ten positions. Three of the ten positions are the decimal positions, and the other seven are positions that are always needed for an e format. Thus, the total width of an e specification must be at least $d+7$; otherwise, asterisks will be printed. The specification above is e10.3.

If there are more decimal positions in the specification than in the exponential form of the value, the extra decimal positions are filled on the right with zeros. If the total width of the e specification is more than seven plus the decimal positions, the extra positions appear as blanks on the left side of the value.

EXAMPLE 2-10

Time in an Exponential Form

Print the value of a real variable named time in an exponential form with four decimal positions.

SOLUTION

Computer Memory

time **−0.00125**

Fortran Statements

```
        print 10, time
    10 format(1x,"time = ",e11.4)
```

Buffer Contents

```
btimeb=b−0.1250E−02
```

Computer Output

```
             111111111
      123456789012345678
```

```
time = −0.1250E−02
```

. . . .

Try It

Try this self-test to check your memory of some key points from Section 2-4. If you have any problems with the exercises, you should reread this section. The solutions are given at the end of this module.

1. What is printed by the following statements? Show the exact location of any blanks.

```
    real :: x
    ...
    x = −27.632
    print 5, x
  5 format(1x,"x = ",f7.1," degrees")
```

2. What is printed by the following statements? Show the exact location of any blanks.

```
    real :: distance=28732.5, velocity
    ...
    velocity = −2.6
    print 5, distance, velocity
  5 format(1x,"distance = ",e10.3,&
          5x,"velocity = ",f5.2)
```

2-5 COMPLETE PROGRAMS

The program statement identifies the beginning of a program and assigns the program name. The general form of this statement is

```
program program_name
```

Like a variable name, the program name can be 1 to 31 characters, must begin with a letter, and contains only letters, digits, and the underscore character. Some example program statements are

```
program homework_1
program compute_density
program sort_data
```

You may not use a variable in your program that has the same name as the program name. The `program` statement is not required, but we recommend using it to clearly identify the beginning of a program.

The `stop` statement signals the computer to terminate execution of the program. It can appear anywhere in the program that makes sense, and it can appear as often as necessary. For example, certain data values may not be valid, and you may want to stop executing the program if they occur. The general form of the `stop` statement is

```
stop
```

The `stop` statement is optional in most programs, as explained below.

The `end` statement identifies the physical end of a Fortran program for the compiler. Since the compiler translates statements until it reaches the end statement, every Fortran program must terminate with the `end` statement. The simplest form of the `end` statement is

```
end
```

but the general form includes a program name:

```
end program program_name
```

Including the program name is optional, but it aids in documenting our programs. In later chapters our programs will contain several procedures, each of which ends with an `end` statement; in these examples, it is especially useful to include the program or procedure name in the `end` statement.

If a `stop` statement does not immediately precede the `end` statement, the compiler automatically adds one. Therefore, a program may use both a `stop` and an `end` at the end of the program, although the `stop` is not necessary. The programs in this module will not include a `stop` statement immediately preceding an `end` statement.

Specification statements must precede *executable statements* in a program. Therefore, simple Fortran programs should have the following general structure:

> program statement
> descriptive comments
> specification statements
> executable statements
> end statement

If you look back at the `compute_average` program in Chapter 1 on page 5, you will be able to identify the various groups of statements.

In addition to learning to write correct programs, it is important to learn to write programs with good programming style, that is, to write programs with a simple and consistent form. The following are some of the style and technique guidelines that we will follow in the example programs in this module:

1. We will mark the beginning and end of each program with a comment line containing a series of dashes. A program statement will always be the first Fortran statement in a program. The end statement for a program will always include the program name.
2. A brief discussion of the purpose of the program will follow the program statement. This information should be similar to the information in the first two steps of our problem-solving process — the problem statement and a description of the input and output.
3. Specification statements will be used to specify the type of all variables used in the program. When possible, variable initializations will be done along with the type specifications.
4. Comment lines containing blanks and additional documentation information will be used to separate programs into sections that correspond to the decomposition of the problem solution.
5. Any statement numbers will be chosen to be in increasing order.

EXAMPLE 2-11

Convert Kilowatt-Hours to Joules

Write a program to convert an amount in kWh (kilowatt-hours) to joules. The amount in kWh is to be entered from the terminal. The output should be the number read in kWh and the converted value in joules. (Use the following conversion factor: joules = 3.6E+06 × kWh.)

SOLUTION

Step 1 is to state the problem clearly:

Convert a value in kilowatt-hours to joules.

Step 2 is to describe the input and output:

Input — value in kilowatt-hours
Output — corresponding value in joules

Step 3 is to work a simple example by hand. Let the input value be 5.5 kilowatt-hours. Then the number of joules is 3.6E+06 × 5.5 = 19.8E+06.

Step 4 is to develop an algorithm. We start with the decomposition and then add more details to obtain the pseudocode. (Pseudocode is discussed more in Chapter 3.) We can usually go directly from the pseudocode to the Fortran program.

Decomposition

Read amount in kilowatt-hours.
Convert amount to joules.
Print both amounts.

Pseudocode

conversion: Read kilowatt-hours
joules ← 3.6E+06 × kilowatt-hours
Print kilowatt-hours, joules

Fortran Program

```
!-------------------------------------------------------------------!
program conversion
!
!  This program reads an input value in
!  kilowatt-hours, and converts it to joules.
!  Both values are then printed.
!
!
!    Define variables.
     implicit none
     real :: joules, kw_hrs
!
!    Prompt the user to enter the energy value.
     print*, "Enter energy in kilowatt-hours:"
     read*, kw_hrs
!
!    Convert energy to joules.
     joules = 3.6e+06*kw_hrs
!
!    Print both energy values.
     print 10, kw_hrs, joules
  10 format(1x,f6.2," kilowatt-hours = ",&
            e9.2," joules")
!
end program conversion
!-------------------------------------------------------------------!
```

Step 5 is to test the program. The following computer output illustrates a sample run of the program using the same data that we used in the hand example:

```
Enter energy in kilowatt-hours:
5.5
  5.50 kilowatt-hours =  0.20E+08 joules
```

In the program in Example 2-11, we printed a message to the user that specified the input that the program was expecting. After converting the value in kilowatt-hours to joules, we printed the value of kilowatt-hours along with the converted value in joules. This interaction between the program and the user resembles a conversation and is called *conversational computing*.

2-6 Application BACTERIA GROWTH

Biomedical Engineering

A biology laboratory experiment involves the analysis of a strain of bacteria. Because the growth of bacteria in the colony can be modeled with an exponential equation, we are going to write a computer program to predict how many bacteria will be in the colony after a specified amount of time. Suppose that, for this type of bacteria, the equation to predict growth is

$$y_{new} = y_{old} \cdot e^{1.386t}$$

where y_{new} is the new number of bacteria in the colony, y_{old} is the initial number of bacteria in the colony, and t is the elapsed time in hours. Thus, when t = 0, we have

$$y_{new} = y_{old} \cdot e^0 = y_{old}$$

The output number of bacteria should always be an integer, even though the equation computes real numbers.

1. Problem Statement

Using the equation

$$y_{new} = y_{old} \cdot e^{1.386t}$$

predict the number of bacteria (y_{new}) in a bacteria colony given the initial number in the colony (y_{old}) and the time elapsed (t) in hours.

2. Input/Output Description

Input—the initial number of bacteria and the time elapsed
Output—the number of bacteria in the colony after the elapsed time

3. Hand Example

You will need your calculator for these calculations. For t = 1 hour and y_{old} = 1 bacterium, the new colony contains

$$y_{new} = 1 \cdot e^{1.386} = 4.00$$

After 6 hours, the size of the colony is

$$y_{new} = 1 \cdot e^{1.386 \times 6} = 4088$$

If we start with 2 bacteria, after 6 hours the size of the colony is

$$y_{new} = 2 \cdot e^{1.386 \times 6} = 8177$$

4. Algorithm Development

Decomposition

Read y_{old}, t.
Compute y_{new}.
Print y_{old}, t, y_{new}.

Pseudocode

growth: Read y_{old}, t

           $y_{new} \leftarrow y_{old} \cdot e^{1.386t}$

           Print y_{old}, t, y_{new}

Fortran Program

```
!-----------------------------------------------------------------|
program growth
!
!   This program reads the initial population and
!   the time elapsed and then computes and prints
!   the predicted population for the bacteria.
!
!
!     Define variables.
      implicit none
      integer :: y_new, y_old
      real :: time
!
!     Prompt the user to enter the initial
!     population and time elapsed.
      print*, "Enter initial population:"
      read*, y_old
      print*, "Enter time elapsed in hours:"
      read*, time
!
!     Compute predicted population.
      y_new = real(y_old)*exp(1.386*time)
!
!     Print population information.
      print 10, y_old
   10 format(1x,"Initial population = ",i4)
```

```
      print 20, time
  20 format(1x,"Time elapsed (hours) = ",f9.4)
      print 30, y_new
  30 format(1x,"Predicted population = ",i12)
!
end program growth
!------------------------------------------------------------------!
```

This program and all other programs from the Application sections are available on a diskette that is available to your instructor. Check with your instructor to see if this information has been stored on your computer system so that you can access these programs.

5. Testing

The program output using data from one of the hand examples is

```
Enter initial population:
1.0
Enter time elapsed in hours:
6.0
Initial population   =  1
Time elapsed (hours) =    6.0000
Predicted population =           4088
```

Since the number of bacteria in a population can become very large over time, you might want to use a real value to store the number of bacteria and then print the number using an e specification. Negative values of time represent population decreases, but if the population falls below 1 the model is no longer applicable.

2-7 Application   STRIDE ESTIMATION

Mechanical Engineering

The physics that govern the motion of a simple pendulum enable us to estimate the time that it takes for a natural stride of a person given only the length of his or her leg. This is done by modeling the leg by a long, thin rod of uniform cross section that pivots about its upper end, as shown in the diagram on page 43.

Using this model, a freely swinging rod supported at its upper end will take T seconds to swing from one end to another and back to its original position. Thus, the formula is[1]

$$T = 2 \cdot \pi \sqrt{\frac{2L}{3g}}$$

where L is the length of the rod (or leg) in feet and g is the acceleration due to gravity, which is approximately 32 ft/s². A stride will take T/2

---

[1] E. R. Jones and R. L. Childer, *Contemporary College Physics* (Reading, Mass.: Addison-Wesley, 1990), pp. 392–393.

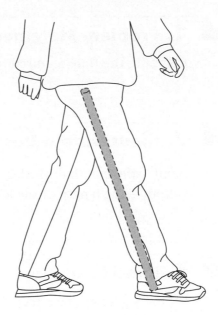

seconds, since a stride is a step forward; it is not a step forward and then back to the original position.

It is interesting to observe that the length of a stride, L_s, is proportional to L, the length of the leg. Thus, we can express L_s as a product of a constant α and L:

$$L_s = \alpha L$$

If we know the length of a stride and the time it takes to take a stride, we can compute the walking speed, which is the number of feet per second traveled while walking:

$$\text{Walking speed} = \frac{L_s}{\dfrac{T}{2}}$$

$$= \frac{\alpha L}{\pi \sqrt{\dfrac{2L}{3g}}}$$

$$= \frac{\alpha}{\pi \sqrt{\dfrac{2}{3g}}} \cdot \sqrt{L}$$

Therefore, the walking speed in feet per second is proportional to the square root of the length of the leg. Thus, people with longer legs have a faster walking speed and go farther per second. This model does not work for running because the leg does not swing freely in a running gait.

Write a program that reads the length of a leg in feet and estimates the time required for a stride.

1. Problem Statement

Compute the time required for a stride.

2. Input/Output Description

Input—the length of the leg
Output—the length of the leg and the time required for a stride

3. Hand Example

Let the leg length be 3 feet. Then T_s, the time for a stride, is

$$T_s = \frac{T}{2} = \pi \sqrt{\frac{2L}{3g}} = \pi \sqrt{\frac{6}{96}} = 0.79s$$

Just for fun, measure your leg (from the hip joint to the heel) and use a stopwatch to time your natural walking stride. Compare that to the value just computed. (It would also be interesting to measure the length of your stride and compute an estimate for the constant α. Compare your estimate for α with estimates from classmates with a wide range of heights. They should be similar.)

4. Algorithm Development

Decomposition

Read leg length.
Compute stride time.
Print stride time.

Pseudocode

stride: Read L

$$T_s \leftarrow \pi \sqrt{\frac{2L}{3g}}$$

Print T_s

The Fortran program that follows defines the constants as separate variables so that the equation for T_s is as close as possible to the original equation. This makes the program easier to read and reduces the possibility of errors in converting the equation into Fortran.

Fortran Program

```
!-------------------------------------------------------------!
program stride
!
!   This program reads the length of a leg, and then
!   computes and prints the time required for a stride.
!
!
!     Define and initialize variables.
      implicit none
      real :: g, L, pi, T_s
      parameter (g=32.0, pi=3.141593)
!
!     Prompt the user to enter the leg length.
      print*, "Enter the leg length in feet:"
      read*, L
!
!     Compute time required for a stride.
      T_s = pi*sqrt(2.0*L/(3.0*g))
!
!     Print stride time and leg length.
      print*
      print 5, T_s
    5 format(1x,"The stride time is ",f5.2," seconds")
      print 10, L
   10 format(1x,"for a leg length of ",f5.2," feet")
!
end program stride
!-------------------------------------------------------------!
```

◢ 5. Testing

If we test this program with the hand example data, the output is the
following:

```
Enter the leg length in feet:
3

The stride time is  0.79 seconds
for a leg length of  3.00 feet
```

2-8 ADDITIONAL FORMATTING FEATURES

This section presents several useful features of format statements. Examples illustrate each feature.

Repetition

If we have two identical specifications in a row, we can use a constant in front of the specification (or sets of specifications) to indicate repetition. For instance, i2,i2,i2 can be replaced by 3i2. Often format statements can be made shorter with repetition constants. The following pairs of format statements illustrate the use of repetition constants:

```
10  format(3x,i2,3x,i2)
10  format(2(3x,i2))

20  format(1x,f4.1,f4.1,1x,i3,1x,i3,1x,i3)
20  format(1x,2f4.1,3(1x,i3))
```

Slash

A slash (/) in a format statement specifies that the current buffer should be printed and a new one started. The slash is especially useful in inserting blank lines in the output. However, do not assume that a slash will always cause single spacing; the spacing following the line printed by the slash depends on the carriage control character of the next line. The slash character may be enclosed in commas if desired.

The following statements print the heading Test Results followed by column headings Time and Height:

```
    print 5
5  format(1x,"Test Results"/1x,"Time",5x,"Height")
```

The column headings are separated by five spaces, as shown:

```
Test Results
Time     Height
```

If we add another slash in the format statement, we have these statements:

```
    print 5
5  format(1x,"Test Results"//1x,"Time",5x,"Height")
```

The execution of these statements gives a blank line between the two headings:

```
Test Results

Time     Height
```

Tab Specification

The tab specification tn allows you to shift directly to a specified position, n, in the output line. The following pairs of format statements function exactly the same:

```
500 format(58x,"Experiment No. 1")
500 format(T59,"Experiment No. 1")

550 format(1x,"sales",10x,"Profit",10x,"Loss")
550 format(1x,"sales",t17,"Profit",t33,"Loss")

600 format(f6.1,15x,i7)
600 format(f6.1,t22,i7)
```

The tln and trn specifications tab left or right for n positions from the current position. The following formats are therefore equivalent:

```
85 format(1x,25x,"Height",5x,"Weight")
85 format(t27,"Height",tr5,"Weight")
```

The tab specifications are especially useful in aligning column headings and data.

Number of Specifications

Suppose there are more format specifications than variables on a print list, as shown:

```
print 5, speed, distance
5 format(4f5.2)
```

In these cases the computer uses as much of the specification list as it needs and ignores the rest. In the example speed and distance would be matched to the first two specifications; the last two specifications would be ignored.

Suppose there are fewer format specifications than variables on a print list, as shown:

```
print 20, temperature, volume
20 format(1x,f6.2)
```

In these cases we match variables and specifications until we reach the end of the format. Then two events occur:

1. We print the current buffer and start a new one.
2. We back up in the format specification list until we reach a left parenthesis, and we again begin matching the remaining variables to the specifications at that point. If a repetition constant is in front of this left parenthesis, it applies to the format specifications being reused.

In the previous statements temperature would be matched to the f6.2 specification. Because there is no specification for volume, we do the following:

1. Print the value of temperature after single-spacing.
2. Back up to the beginning of the format specification list (first left parenthesis), and match the f6.2 to the value of volume. We then reach the end of the list and single-space to print the value of volume. Thus, temperature and volume are printed on separate lines.

This discussion allows you to understand what happens if the number of specifications is not the same as the number of variables. However, we recommend that you reference the same format with more than one print statement only when the format specifications are exactly the same. Write a new format statement if the number of variables is different; this keeps the program simpler to understand.

Try It

Try this self-test to check your memory of some key points from Section 2-8. If you have any problems with the exercises, you should reread this section. The solutions are given at the end of this module.

In problems 1–3, show the output from the following print statements. Be sure to indicate the vertical as well as the horizontal spacing. Use the following real variables and corresponding values:

time	3.5
response__1	178.8
response__2	0.00204

1. `print 5, time, response_1, response_2`
    `5 format(1x,f6.2,t20,2f7.4)`
2. `print 15, time, response_1, response_2`
    `15 format(1x,"time = ",f5.2,tr5," response 1 = ",e9.2/&`
    `         1x,"time = ",f5.2,tr5," response 2 = ",e9.2)`
3. `print 20, time, response_1, response_2`
    `20 format(1x,"Experiment Results"//&`
    `         1x,"Time",2x,"Response 1",2x,"Response 2"/&`
    `         1x,f4.2,2f12.3)`

In problems 4–7 tell how many data lines are printed by the following print statements. Indicate which variables are on each line and which columns are used.

4. `print 30, time, distance, velocity, acceleration`
    `30 format(4(1x,f6.3))`
5. `print 40, time, distance, velocity, acceleration`
    `40 format(1x,f6.2)`
6. `print 25, time, distance, velocity, acceleration`
    `25 format(1x,f4.2/1x,f4.1)`
7. `print 20, time, distance`
    `print 20, velocity, acceleration`
    `20 format(1x,4f6.3)`

SUMMARY

This chapter discussed how to define variables and constants in Fortran. It also discussed the arithmetic operations and intrinsic functions that allow us to compute new values using these variables and constants. Some of the considerations that are unique to computer computations were discussed with specific examples: magnitude limitations, truncation, mixed-mode operations, underflow, and overflow. Statements for reading information from the terminal and for printing answers were covered. Several complete example programs were developed.

Key Words

argument	intermediate result
assignment statement	intrinsic function
buffer	left-justified
carriage control character	list-directed I/O
comment	literal
constant	mixed-mode operation
continuation	nonexecutable statement
conversational computing	overflow
executable statement	real value
explicit typing	right-justified
exponential notation	rounding
fixed-source form	scientific notation
formatted I/O	specification
free-source form	statement number
generic function	truncation
implicit typing	type statement
initialize	underflow
integer value	variable

Problems

This problem set begins with modifications to programs given earlier in this chapter. Give the decomposition, pseudocode, and Fortran program for each problem.

Problem 1 modifies the energy conversion program conversion given in Section 2-5 on page 39.

1. Modify the energy conversion program so that it converts kilowatt-hours to calories (1 calorie equals 4.19 joules).

Problem 2 modifies the bacteria growth program growth given in Section 2-6 on page 41.

2. Modify the bacteria growth program so that the program reads two time values from the terminal, with no restrictions on which time is larger. Compute and print the amount of bacteria growth that occurs between the two times, assuming an initial population value of 1. (Hint: Review the absolute value function.)

Problem 3 modifies the stride estimation program stride given in Section 2-7 on page 45.

3. Experimentally determine a value for the constant α. Then modify the stride estimation program so that it accepts the leg length and computes the time required for a stride, the length of a stride, and the walking speed.

Develop programs for problems 4–8. Use the five-step problem-solving process.

4. Hot-air balloons ascend because heated air is less dense than cool air. As the air inside is heated, it expands, and some escapes in order to maintain constant atmospheric pressure. The mass of air inside the balloon, for a fixed volume, is less than the mass of an equivalent volume of cooler air outside. Write a program to determine the mass of air that remains (m_2) after reading the original mass of air (m_1), the balloon volume (v_1), and the total volume of the air after heating (v_2). Use the following equation for computing m_2:

$$m_2 = \frac{v_1}{v_2} m_1$$

5. When a train travels over a straight section of track, it exerts a downward force on the rails; but when it rounds a level curve, it also exerts a horizontal force outward on the rails. Both of these forces must be considered when designing the track. The downward force is equivalent to the weight of the train. The horizontal force, called centrifugal force, is a function of the weight of the train, its speed as it rounds the curve, and the radius of the curve. The equation for the horizontal force in pounds is

$$force = \frac{weight \cdot 2000}{32} \cdot \frac{(mph \cdot 1.4667)^2}{radius}$$

where weight is the weight of the train in tons, mph is the speed of the train in miles per hour, and radius is the radius of the curve in feet. Write a program to read values for weight, mph, and radius. Compute and print the corresponding horizontal force.

6. Modify the program in problem 5 so that the speed is entered in kilometers per hour instead of miles per hour. (Recall that 1 mile is equal to 1.609 kilometers.)

7. A research scientist performed nutrition tests using three animals. Data on each animal include an identification number, the weight of the animal at the beginning of the experiment, and the weight of the animal at the end of the experiment. Write a program to read this data and print a report. The report is to include the original information plus the percentage increase in weight for each animal.

8. Write a program to read the following information from the terminal:

> Year
>
> Number of people in the civilian labor force
>
> Number of people in the military labor force

Compute the percentage of the labor force that is civilian and the percentage that is military. Print the following information:

```
Labor Force - Year xxxx
Number of Workers (thousands) and Percentage of Workers
Civilian          xxx.xxx     xxx.xxx
Military          xxx.xxx     xxx.xxx
Total             xxx.xxx     xxx.xxx
```

3 Control Constructs

An optical fiber (a transparent glass thread) is thinner than a human hair, but it can carry more information than either radio waves or electrical waves in copper telephone wires. In addition, fiber optic communication signals do not produce electromagnetic waves that cause "crosstalk" noise on communication lines. The first transoceanic fiber optic cable was laid in 1988 across the Atlantic. It contains four fibers that can handle up to 40,000 calls at one time. Other applications for optical fibers include motion sensing in gyroscopes, linking industrial lasers to machining tools, and threading light into the human body for examinations and laser surgery.

In Chapter 2 we wrote complete Fortran programs, but the steps were all executed sequentially. The programs read data, computed new data, and printed the new data. This chapter introduces Fortran statements that allow us to control the sequence of the steps being executed. This control is achieved through statements that allow us to select different paths through our programs and statements that allow us to repeat certain parts of our programs. These new statements are used to implement *control structures,* or control constructs.

3-1 ALGORITHM STRUCTURE

Chapter 1 presented the following five-step process for developing problem solutions:

1. State the problem clearly.
2. Describe the input and output information.
3. Work the problem by hand (or with a calculator) for a simple set of data.
4. Develop a solution, or algorithm, that is general in nature.
5. Test the algorithm with a variety of data sets.

To describe algorithms consistently, we use a set of standard forms, or structures. When an algorithm is described in these standard structures, it is a *structured algorithm.* When the algorithm is converted into computer instructions, the corresponding program is a *structured program.* This chapter begins with a discussion of pseudocode and flowcharts, which describe the steps in an algorithm; we then discuss organizing the steps in standard structures.

Pseudocode and Flowcharts

Each element for building algorithms can be described in an English-like notation called *pseudocode.* Because pseudocode is not really computer code, it is language independent; that is, pseudocode depends only on the steps needed to solve a problem, not on the computer language that will be used for writing the solution. *Flowcharts* also describe the steps in algorithms, but they are a graphical description, as opposed to a set of English-like steps. Neither pseudocode nor flowcharts are intended to be a formal way of describing the algorithm; they are informal ways of easily describing the steps in the algorithm without worrying about the syntax of a specific computer language. This chapter includes a number of examples of both pseudocode and flowcharts so that you can compare the techniques and choose the one that works best for you.

The basic operations that we perform in algorithms are computations, input, output, and comparisons. Table 3-1 compares examples of pseudo-code notation and flowchart symbols for these basic operations, plus the

Table 3-1   Pseudocode Notation and Flowchart Symbols

Basic Operation	Pseudocode Notation	Flowchart Symbol
Computations	average $\leftarrow \dfrac{sum}{count}$	$average = \dfrac{sum}{count}$
Input	read a, b	Read a, b
Output	print a, b	Print a, b
Comparisons	If b > 0.0 then . . .	Is b > 0.0? yes / no
Beginning of algorithm	report:	Start Report
End of algorithm		Stop Report

notation and symbols for identifying the beginning and end of an algorithm. Do not worry about whether the names you have chosen are valid Fortran names; both pseudocode and flowcharts are describing the operations, and they are independent of the language that will be used to implement the computer solution.

In pseudocode a computation is indicated with an arrow, as shown in the following statement:

$$\text{average} \leftarrow \frac{\text{sum}}{\text{count}}$$

We read this pseudocode statement as "the average is replaced by the sum divided by the count" or "the sum is divided by the count, giving the average."

We use comparisons to ask a question within an algorithm. Then, depending on the answer to the question, we can perform one set of statements as opposed to another set of statements. The questions that we ask must be answered with yes or no (or true or false). For example, we can determine if a data value b is greater than zero by using the expressions "If b > 0.0, then . . ." or "Is b > 0.0?"

An algorithm defined in pseudocode begins with the name of the algorithm, followed by the first step in the algorithm. Thus, if the first step in an algorithm named report is to set a sum to zero, the first pseudocode statement is

$$\text{report:} \qquad \text{sum} \leftarrow 0$$

Pseudocode does not need a special statement to signify the end of the algorithm; we assume that the end occurs after the last step described in pseudocode.

General Structures

The steps in an algorithm can be divided into three general structures: sequences, selections, and repetitions. A *sequence* is a set of steps that are performed sequentially, or one after another. A *selection* structure allows us to compare two values and then specify a different set of steps to execute, depending on the result of the comparison. We use the *repetition* structure when we want to repeat a set of steps. We can repeat the steps a specified number of times, or we can repeat the steps as long as a specified condition is true. Now let us look at each of these structures separately.

Sequence   A sequence is a set of steps in an algorithm that are performed sequentially. For example, all of the programs in Chapter 2 were sequential algorithms that contained an input step, a computation step, and an output step. We included sample pseudocode in the development of the programs in Chapter 2; shown on the next page is the flowchart for the program growth from Section 2-6.

Selection   In the selection structure, a comparison is performed to determine which steps are to be performed next. The selection structure is commonly described in terms of an *if structure* that can have several forms.

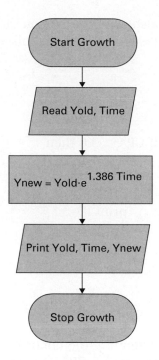

All forms of the if structure include a *condition* that can be evaluated to be true or false. If the condition is true, then a step or a group of steps is performed. For example, suppose the algorithm that we are developing must count the number of students who are on the honor roll and print their student ID numbers. One step in the algorithm might be "if the grade point average (GPA) is greater than or equal to 3.0, then print the ID and add 1 to the number of students on the honor roll." This step is shown in both pseudocode and a flowchart; note the use of indenting in the pseudocode.

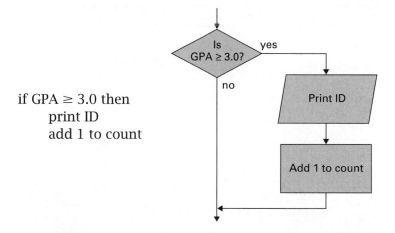

if GPA ≥ 3.0 then
    print ID
    add 1 to count

Another form of the if structure contains an additional clause, called an else clause, that specifies an alternate set of steps to perform if the condition is false. For example, suppose we want to modify the example so that we print the ID of each student. Then, if the student is on the honor roll, we also print the GPA on the same line beside the student's ID. These steps are described with the following pseudocode and flowchart:

if GPA ≥ 3.0 then
    print ID, GPA
    add 1 to count
else
    print ID

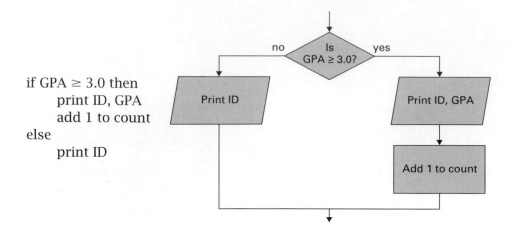

The last form of the if structure contains else if clauses that allow us to test for multiple conditions. A single else if clause can be illustrated by extending the honor roll example. Assume that the president's honor roll requires a GPA greater than 3.5 and that the dean's honor roll requires a GPA greater than or equal to 3.0. We want to count the number of students on each honor roll. We also want to print each student's ID and GPA if the GPA is equal to or above 3.0; in addition, we want to include an asterisk beside the GPA if it is above 3.5. The following pseudocode and flowchart describe this set of steps:

if GPA > 3.5 then
    print ID, GPA, '*'
    add 1 to president's count
    add 1 to dean's count
else if GPA ≥ 3.0 then
    print ID, GPA
    add 1 to dean's count
else
    print ID

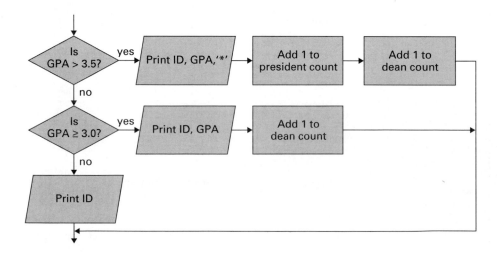

Repetition   A repetition structure allows us to use *loops,* which are sets of steps in an algorithm that are repeated. One type of loop, called a *while loop,* repeats the steps as long as a certain condition is true. Another type of loop, called a *counting loop,* repeats the steps a specified number of times. The steps within a loop can contain sequential steps, selection steps, or other repetition steps.

Suppose that we want to continue reading data values and adding them to a sum as long as that sum is less than 1000. These steps are described in the following pseudocode and flowchart:

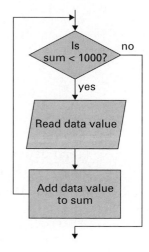

while sum < 1000 do
    read data value
    add data value to sum

Again, note that the indenting in the pseudocode specifies the steps that are included in the while loop. If we wish to print the value of the sum after exiting the while loop, we can use the following pseudocode and flowchart:

while sum < 1000 do
    read data value
    add data value to sum
print sum

Counting loops repeat a set of steps a specified number of times. For example, suppose we are going to read ten values from the terminal and perform some calculations with each value. We can consider a counting loop to be a special form of a while loop in which a counter has been introduced. The counter represents the number of times that the loop has been executed. The counter is usually initialized to zero before the loop is executed. Inside the loop, the counter is incremented by one. The loop is then executed while the value of the counter is less than a specified upper limit. The pseudocode and flowchart for this counting loop example are the following:

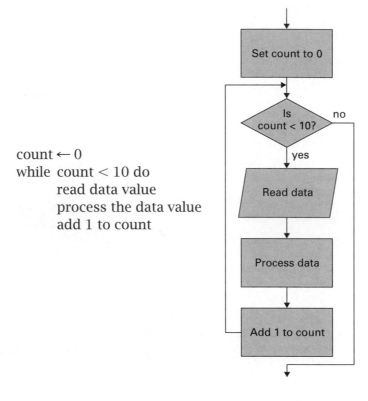

count ← 0
while count < 10 do
      read data value
      process the data value
      add 1 to count

We can describe algorithms for almost any problem using these three structures: sequential steps, selection steps, and repetitive steps.

3-2 IF STRUCTURES

Section 3-1 presented the pseudocode and flowcharts for the three forms of the selection structure; this section presents the corresponding Fortran statements. All three forms use an if structure, which contains a logical expression that is evaluated to determine which path to take in the structure. Therefore, before discussing the Fortran statements, this section covers logical expressions.

Logical Expressions

A *logical expression* is analogous to an arithmetic expression but is always evaluated to either true or false instead of a number. Logical expressions can be formed using the following *relational operators* (in either character or symbol form):

Relational Operator Forms		**Interpretation**
Character	**Symbol**	
.eq.	==	Equal to
.ne.	/=	Not equal to
.lt.	<	Less than
.le.	<=	Less than or equal to
.gt.	>	Greater than
.ge.	>=	Greater than or equal to

Numeric variables can be used on both sides of the relational operators to yield a logical expression whose value is either true or false. For example, consider the logical expression a == b, where a and b are real values. If the value of a is equal to the value of b, then the logical expression a == b is true; otherwise, the expression is false. Similarly, if the value of x is 4.5, then the expression x > 3.0 is true.

We can also combine two logical expressions into a *compound logical expression* with the *logical operators* .or. and .and.. When two logical expressions are joined by .or., the entire expression is true if either or both expressions are true; it is false only when both expressions are false. When two logical expressions are joined by .and., the entire expression is true only if both expressions are true. These logical operators are used only between complete logical expressions. For example, (a < b).or.(a < c) is a valid compound logical expression because .or. joins a < b and a < c. However, (a < b).or.c is an invalid compound expression because c is a numeric variable, not a complete logical expression. Note that the parentheses in these expressions are not required but serve to make the expressions more readable.

Logical expressions can also be preceded by the .not. logical operator. This operator changes the value of the expression to the opposite value; hence, if a > b is true, then .not.(a > b) is false.

A logical expression may contain several logical operators, as in

$$.not.(a < 15.4).or.(count == sum)$$

The *hierarchy* of logical expressions, from highest to lowest, is .not., .and., and .or.. In the preceding statement the logical expression a < 15.4 would be evaluated, and its value, true or false, would then be reversed. This resultant value would be used, along with the value of count == sum, with the logical operator .or.. For example, if a is 5.0, count is 5, and sum is 5, then the left-hand expression is false and the right-hand expression is true; since these expressions are connected by .or., the entire expression is true.

Another type of variable, a *logical variable,* is also useful in writing the logical expressions that allow us to choose different paths and to repeat parts of our programs. A logical variable can have one of two values: true or false. A logical variable must be defined with a specification statement whose form is

> logical *variable list*

Logical constants are `.true.` and `.false.`. Therefore, the statement necessary to define a logical variable and give it a value of false is

```
logical :: done=.false.
```

Logical variables are generally used to make programs more readable; thus, they are not usually part of the input or output. Several examples in this chapter will compare solutions using logical variables to ones without logical variables.

Logical variables cannot be compared with the relational operators presented; instead, two logical variables must be compared with the relations `.eqv.` or `.neqv.`, which represent equivalent and not equivalent. To compare `.not.done` to the value `.true.`, we use this statement:

```
if (.not.done.eqv..true.) then
   ...
```

Table 3-2 summarizes the evaluation of the logical operators for all possible cases.

Whenever arithmetic, relational, and logical operators are in the same expression, the arithmetic operations are performed first. The relational operators are then applied to yield true or false values, and these values are evaluated with the logical operators, whose precedence is `.not.`, `.and.`, and `.or.`. The relations `.eqv.` and `.neqv.` are evaluated last.

As an example, assume that `error` is a logical variable and that a, b, and c are real variables. Consider the following expression:

```
(a > b+c).or.error
```

This expression will be true if either a > b+c or error is `.true.`; the expression will also be true if both a > b+c and error are `.true.`. Table 3-3 lists the precedence of operators in a logical expression.

Logical `if` Statement

Fortran implements a logical `if` statement and a block `if` statement. The logical `if` statement is used when a single statement is to be performed if a logical expression is true. We use the block `if` statement when we want to perform several statements if a logical expression is true, or when we want to use an `else` statement, or when we want to use an `else if` statement.

Table 3-2 Logical Operators

a	b	.not.a	a.and.b	a.or.b	a.eqv.b	a.neqv.B
false	false	true	false	false	true	false
false	true	true	false	true	false	true
true	false	false	false	true	false	true
true	true	false	true	true	true	false

Table 3-3   Relational and Arithmetic Operator Precedence

Precedence	Operator	Associativity
1	Parentheses ( )	Innermost first
2	Exponentiation	Right to left
3	Binary operators * /	Left to right
4	Unary operators + –	Left to right
5	Binary Operators + –	Left to right
6	Relational operators	Left to right
7	.not.	Left to right
8	.and.	Left to right
9	.or.	Left to right
10	.eqv., .neqv.	Left to right

The general form of the logical `if` statement is

> `if` *(logical expression) executable statement*

Execution of this statement consists of the following steps:

1. If the logical expression is true, we execute the statement that is on the same line as the logical expression and then go to the next statement in the program.
2. If the logical expression is false, we jump immediately to the next statement in the program.

A typical `if` statement is the following:

```
if (a < b) sum - sum + a
```

If the value of a is less than the value of b, then the value of a is added to sum. If the value of a is greater than or equal to b, then control passes to whatever statement follows the `if` statement in the program. Other examples of `if` statements are

```
if (time > 1.5) read*, distance
if (denominator <= 0.0) print*, denominator
if (number /= -4) number = number + 1
```

The executable statement that follows the logical expression is typically a computation or an input/output statement — it cannot be another `if` statement.

Block `if` Statement

In many instances we would like to perform more than one statement if a logical expression is true. The form of the `if` statement that allows us to perform any number of statements if a logical expression is true uses the words then and end `if` to identify these steps. The general form is

```
if (logical expression) then
    statement 1
    ...
    statement n
end if
```

Execution of this block of statements consists of the following steps:

1. If the logical expression is true, we execute statements 1 through n and then go to the statement following `end if`.
2. If the logical expression is false, we jump immediately to the statement following `end if`.

Although not required, indenting the statements to be performed when the logical expression is true will indicate that they are a group of statements within the `if` statement. Also, any of the statements 1,2,..., n can be other `if` statements.

EXAMPLE 3-1

Zero Divide

Assume that you have calculated the numerator and the denominator of a fraction. Before dividing the two values, you want to see if the denominator is close to zero. If the denominator is close to zero, you want to print an error message and stop the program; if the denominator is not close to zero, you want to compute the result and print it. Write the statements to perform these steps.

SOLUTION

```
real :: denominator, fraction, numerator
...
if (abs(denominator) < 0.1e-06) then
    print*, "denominator is close to zero"
    stop
end if
fraction = numerator/denominator
```

. . . .

else Statement

The `else` statement allows us to execute one set of statements if a logical expression is true and a different set if the logical expression is false. The general form of an `if` statement combined with an `else` statement (an `if-else` statement) is shown next. If the logical expression is true, then statements 1 through n are executed. If the logical expression is false, then statements n+1 through m are executed. Any statement can also be another `if` or `if-else` statement to provide a nested structure.

```
if (logical expression) then
    statement 1
    ...
    statement n
else
    statement n+1
    ...
    statement m
end if
```

Now consider this set of statements that uses a logical variable `valid` with an `if` structure:

```
real :: volts
logical :: valid
...
if ((-5.0 <= volts).and.(volts <= 5.0)) then
    valid = .true.
else
    valid = .false.
end if
if (.not.valid) print*, "error in data"
```

In these statements the variable `valid` is true if the value in `volts` is between -5.0 and 5.0 (including the values -5.0 and 5.0); otherwise, `valid` is false. After determining the value of `valid`, it can be used in other places in the program, such as in an `if` statement to print an error message if the value is not valid. The advantage of using the logical variable is that we can reference it each time we need to know if the value is valid, instead of repeating the test to see if the value in `volts` is between -5.0 and 5.0 (including the values -5.0 and 5.0).

`else if` Statement

When we nest several levels of `if-else` statements, it may be difficult to determine which logical expressions must be true (or false) to execute a set of statements. In these cases we can use the `else if` statement to clarify the program logic. The general form of this statement is

```
if (logical expression 1) then
    statement 1
    ...
    statement m
else if (logical expression 2) then
    statement m+1
    ...
    statement n
else if (logical expression 3) then
    statement n+1
    ...
    statement p
else
    statement p+1
    ...
    statement q
end if
```

This example has two `else if` statements; an actual construction may have more or less. If logical expression 1 is true, then only statements 1 through m are executed. If logical expression 1 is false and logical expression 2 is true, then only statements m+1 through n are executed. If logical expressions 1 and 2 are false and logical expression 3 is true, then only statements n+1 through p are executed. If more than one logical expression is true, the first true logical expression encountered is the only one executed.

If none of the logical expressions are true, then statements p+1 through q are executed. If there is not a final `else` statement and none of the logical expressions are true, then the entire construction is skipped.

EXAMPLE 3-2   ## Weight Category

An analysis of a group of weight measurements involves converting a weight value into an integer category number that is determined as follows:

Category	Weight (pounds)
1	weight ≤ 50.0
2	$50 <$ weight ≤ 125.0
3	$125.0 <$ weight ≤ 200.0
4	$200.0 <$ weight

Write Fortran statements that put the correct value (1, 2, 3, or 4) into a variable named `category` based on the value of a real variable `weight`.

SOLUTION 1

This solution uses nested `if-else` statements:

```
integer :: category
real :: weight
...
if (weight <= 50.0) then
   category = 1
else
   if (weight <= 125.0) then
      category = 2
   else
      if (weight <= 200.0) then
         category = 3
      else
         category = 4
      end if
   end if
end if
```

SOLUTION 2

This solution uses the `if-else if` statement:

```
if (weight <= 50.0) then
    category = 1
else if (weight <= 125.0) then
    category = 2
else if (weight <= 200.0) then
    category = 3
else
    category = 4
end if
```

As you can see, Solution 2 is more compact than Solution 1: It combines the `else` and `if` statements into the single statement `else if` and eliminates two of the `end if` statements.

   The order of the logical expressions is important in these two solutions because the evaluation will stop as soon as a true logical expression has been encountered. Changing the order of the logical expressions and the category assignments can cause the `category` value to be set incorrectly.

. . . .

Try It

Try this self-test to check your memory of some key points from Section 3-2. If you have any problems with the exercises, you should reread this section. The solutions are given at the end of this module.

For problems 1–8 use the values given to determine whether the following logical expressions are true or false. Assume that a and b are real variables, i is an integer variable, and done is a logical variable.

a `2.2`   b `−1.2`
i `5`   done `.true.`

1. a < b

2. a–b >= 6.5

3. i /= 5

4. a+b >= b

5. .not.(a == 2*b)

6. i <= i−5

7. (a < 10.0).and.(b > 5.0)

8. (abs(i) > 2).or.done

For problems 9–16, give Fortran statements that perform the steps indicated. Assume that all the variables used in these steps are real variables.

  9. If `time` is greater than 5.0, increment `time` by 0.5.

10. When the square root of `polynomial` is greater than or equal to 8.0, print the value of `polynomial`.

11. If the difference between `volt_1` and `volt_2` is smaller than 6.0, print the values of `volt_1` and `volt_2`.

12. If the absolute value of `denominator` is less than 0.005, print the message `Denominator is too small`.

13. If the natural logarithm of `x2` is greater than or equal to 3, set `time` equal to zero and add `x` to `sum`.

14. If `distance` is less than 50.0 or `time` is greater than 10.0, increment `time` by 1.0. Otherwise, increment `time` by 0.5.

15. If distance is greater than or equal to 50.0, increment time by 2.0 and print a message Distance > 50.0.

16. If distance is greater than 100.0, increment time by 2.0. If distance is between 50.0 and 100.0 (including 100.0), increment time by 1.0; otherwise, increment time by 0.5.

| 3-3 Application | **LIGHT PIPES** |

Optical Engineering

If we direct light into one end of a long rod of glass or plastic, the light is totally reflected by the walls, bouncing back and forth until it emerges at the far end of the rod. Light pipes use this interesting optical phenomenon to transmit light, and even images, from one place to another. If we bend a light pipe, the light will follow the shape of the pipe and emerge only at the end, as shown in the diagram below:

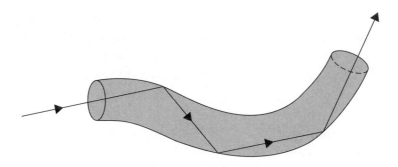

Light pipes made of very thin optical fibers can be grouped into bundles. If the fiber ends are polished and the spatial arrangement is the same at both ends (a coherent bundle), the fiber bundle can be used to transmit an image, and the bundle is called an image conduit. If the fibers do not have the same arrangement at both ends (an incoherent bundle), light is transmitted instead of an image, and the bundle is called a light guide. Because the optical fibers are so flexible, light guides and image conduits are used in instruments designed to permit visual observation of objects or areas that would otherwise be inaccessible. For example, an endoscope is an instrument used by physicians to examine the interior of a patient's body with only a very small incision.

This phenomenon of total internal reflection can be predicted using Snell's law and the indices of refraction of the materials being considered for the light pipe. A light pipe is actually composed of two materials: the material forming the rod or pipe itself and the material that surrounds the pipe. Normally, the material forming the rod is denser than the surrounding medium. When light passes from one material into another material with a different density, the light is bent, or refracted, at the interface of the two materials. The amount of refraction depends on the indices of refraction of the materials and the angle of incidence of the light. If the light striking the interface comes from within the denser material, it may reflect off the interface rather than pass through it. The angle of incidence where the light will be reflected from the surface, rather than cross it, is called the critical angle θ_c. Since the critical angle depends on the indices of refraction of the two materials, we can compute this angle and determine whether light entering the pipe at a particular angle will stay within the pipe. Assume n_2 is the index of refraction of the surrounding medium and n_1 is the index of refraction of the pipe itself. If n_2 is greater than n_1, the pipe will not transmit light; otherwise, the critical angle can be determined from the following equation:

$$\sin \theta_c = \frac{n_2}{n_1}$$

Write a program that reads the indices of refraction for two materials that form a pipe and the angle at which light enters the pipe. Print a message indicating if the pipe will transmit light or not.

1. Problem Statement

Determine whether a light pipe generated from two materials will transmit light that enters it at a given angle.

2. Input/Output Description

Input—indices of refraction of the two materials and the angle at which light enters the pipe

Output—message indicating whether or not light is transmitted

3. Hand Example

The index of refraction of air is 1.0003, and the index of glass is 1.5. If we form a light pipe of glass surrounded by air, the critical angle θ_c can be computed as follows:

$$\theta_c = \sin^{-1}\left(\frac{n_2}{n_1}\right)$$
$$= \sin^{-1}\left(\frac{1.0003}{1.5}\right)$$
$$= \sin^{-1}(0.66687)$$
$$= 41.82°$$

This light pipe will transmit light for all angles of incidence greater than 41.82°.

4. Algorithm Development

Decomposition

Read n_1, n_2, and incidence angle.
Determine appropriate message.

We will need to make sure that n_2 is not greater than n_1 before we compute the critical angle, since the inverse sine function will give an error message if its argument is greater than 1.0.

Pseudocode

light_pipe: read n_1, n_2, and incidence angle
                  if $n_2 > n_1$, then
                        print "Light is not transmitted."
                  else

$$\text{critical angle} \leftarrow \sin^{-1}\left(\frac{n_2}{n_1}\right)$$

                    if incidence angle > critical angle then
                        print "Light is transmitted."
                    else
                        print "Light is not transmitted."

As we convert the pseudocode into Fortran, we need to be sure to remind the user to use the proper units for the angle measurement. We can write the program to accept either radians or degrees, but the equation for computing the critical angle will depend on the units of the input angle.

Fortran Program

```
!-----------------------------------------------------------------!
program light_pipe
!
!   This program reads the indices of refraction for two
!   materials forming a light pipe. It also reads the
!   angle of incidence for light striking the pipe and
!   determines if the light is transmitted.
!
!
!     Define and initialize variables.
      implicit none
      real :: angle, critical_angle, n_1, n_2, pi
      parameter (pi=3.141593)
!
!     Prompt user to enter indices of refraction.
      print*, "Enter index of refraction for rod:"
      read*, n_1
      print*, "Enter index of refraction", &
             " for surrounding medium:"
      read*, n_2
```

```
        print*, "Enter angle of transmission of", &
               " light in degrees:"
        read*, angle
!
!       Determine if light is transmitted and print
!       an appropriate message.
        if (n_2 > n_1) then
            print*, "Light is not transmitted."
        else
            critical_angle = asin(n_2/n_1)*(180.0/pi)
            if (angle > critical_angle) then
               print*, "Light is transmitted."
            else
               print*, "Light is not transmitted."
            end if
        end if
!
end program light_pipe
!-------------------------------------------------------------------!
```

◢ 5. Testing

To test this program with a variety of possible pipes, we can use the following list of indices of refraction for some common materials.[1]

Material	Index of Refraction
Gases (at atmospheric pressure and 0° C)	
Hydrogen	1.0001
Air	1.0003
Carbon dioxide	1.0005
Liquids (at 20° C)	
Water	1.333
Ethyl alcohol	1.362
Glycerine	1.473
Solids (at room temperature)	
Ice	1.31
Polystyrene	1.59
Crown glass	1.50 – 1.62
Flint glass	1.57 – 1.75
Diamond	2.417
Acrylic (polymethylmethacrylate)	1.49

---

[1] E. R. Jones and R. L. Childers, *Contemporary College Physics* (Reading, Mass.: Addison-Wesley, 1990).

If we use this program to determine whether light with an incidence angle of 45° is transmitted through an acrylic pipe in water, we get the following output:

```
Enter index of refraction for rod:
1.49
Enter index of refraction for surrounding medium:
1.333
Enter angle of transmission of light in degrees:
45.0
Light is not transmitted.
```

3-4 WHILE LOOP STRUCTURE

The while loop is an important structure for repeating a set of statements as long as a certain condition is true. In pseudocode, the while loop structure is

```
while condition  do
    statement 1
    ...
    statement m
statement p
```

While the condition is true, statements 1 through m are executed. After the group of statements is executed, the condition is retested. If the condition is still true, the group of statements is reexecuted. When the condition is false, execution continues with the statement following the while loop (statement p in our example). The variables modified in the group of statements in the while loop must involve the variables tested in the while loop's condition, or the value of the condition will never change.

The while loop is implemented in Fortran 90 with the do while statement, which has the following general structure:

```
do while (logical expression)
    statement 1
    ...
    statement m
end do
```

When this structure is executed, the logical expression is evaluated, and if it is true, then statements 1 through m are executed. Control returns to the do while statement, the logical expression is again evaluated, and, if it is still true, statements 1 through m are again executed. This repetition continues until the logical expression is false.

EXAMPLE 3-3

Average of a Set of Data Values

The flowchart shown here was developed for an algorithm to find the average of a set of data values. The data values are assumed to be positive; a negative value indicates the end of the data. Convert the flowchart into a program.

SOLUTION 1

We have covered the Fortran statements for the if structures and the while loop. Using these statements, we can translate each step in the flowchart into Fortran.

```
!------------------------------------------------------------------!
program compute_average_1
!
!   This program computes and prints the average of
!   a set of experimental data values that are all
!   nonnegative. A negative value indicates the
!   end of the experimental data values.
!
!
!       Define and initialize variables.
        implicit none
        integer :: count=0
        real :: average, sum=0.0, x
!
!       Prompt the user to enter the data.
        print*, "Enter data values, one per line."
        print*, "(Use a negative value for end of data.)"
!
!       While not at the end of the data, update sum.
        read*, x
        do while (x >= 0.0)
           sum = sum + x
           count = count + 1
           read*, x
        end do
!
!       Compute average.
        average = sum/real(count)
!
!       Print average.
        print 5, average
    5 format(1x,"The average is ",f5.2)
!
end program compute_average_1
!------------------------------------------------------------------!
```

SOLUTION 2

We now present a second solution that uses a logical variable done to specify when we have reached the end of the data. In this solution, we initialize the

Flowchart:

- Start Compute
- Set sum and count to zero
- Read data value
- Is data value ≥ 0? — no
- yes
- Add data value to sum. Add 1 to count.
- Read data value
- $average = \dfrac{sum}{count}$
- Print average
- Stop Compute

logical variable to the value .false.; when we find the value that indicates the end of the data, we change the value of the logical variable to .true..

```fortran
!------------------------------------------------------------!
program compute_average_2
!
!  This program computes and prints the average of
!  a set of experimental data values that are all
!  nonnegative. A negative value indicates the
!  end of the experimental data values.
!
!
!     Define and initialize variables.
      implicit none
      integer :: count=0
      real :: average, sum=0.0, x
      logical :: done=.false.
!
!     Prompt the user to enter the data.
      print*, "Enter data values, one per line."
      print*, "(Use a negative value for end of data.)"
!
!     While not at the end of the data, update sum.
      do while (.not.done)
         read*, x
         if (x >= 0.0) then
            sum = sum + x
            count = count + 1
         else
            done = .true.
         end if
      end do
!
!     Compute average.
      average = sum/real(count)
!
!     Print average.
      print 5, average
    5 format(1x,"The average is ",f5.2)
!
end program compute_average_2
!------------------------------------------------------------!
```

Note that both of these solutions assume that at least one positive value is entered; otherwise, the variable count is equal to zero and a division by zero occurs.

. . . .

3-5 Application   ROCKET TRAJECTORY

Aerospace Engineering

A small rocket is being designed to make wind shear measurements in the vicinity of thunderstorms. Before testing begins the designers are developing a simulation of the rocket's trajectory. They have derived the following equation, which they believe will predict the performance of their test rocket, where t is the elapsed time in seconds and height is computed in feet:

$$\text{height} = 60 + 2.13\,t^2 - 0.0013\,t^4 + 0.000034\,t^{4.751}$$

The equation gives the height above ground level at time t. The first term (60) is the height in feet above ground level of the nose of the rocket. To check the predicted performance, the rocket is "flown" on a computer, using the preceding equation.

   Develop an algorithm and use it to write a complete program to cover a maximum flight of 100 seconds. Increments in time are to be 2.0 seconds from launch through the ascending and descending portions of the trajectory until the rocket descends to within 50 feet of ground level. Below 50 feet the time increments are to be 0.05 seconds. If the rocket impacts prior to 100 seconds, the program is to stop immediately after impact. The output is to be a table of corresponding time and height values.

   As shown in the following diagram, several possible events could occur as we simulate the flight. The height above the ground should increase for a period and then decrease until the rocket impacts. We can test for impact by testing the height for a value equal to or less than zero. It is also possible that the rocket will still be airborne after 100 seconds of flight time. Therefore, we must also test for this condition and stop the program if the value of time becomes greater than 100. In addition, we need to observe the height above ground. As the rocket approaches the ground, we want to monitor its progress more frequently; we will reduce the time increment from 2.0 seconds to 0.05 seconds.

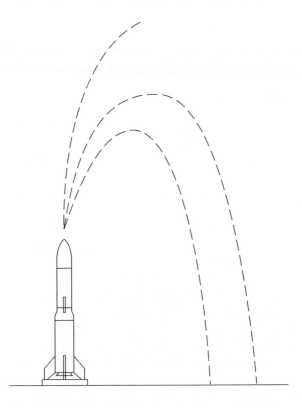

1. Problem Statement

Print a table of time and height values for a rocket trajectory. Start time at zero, and increment it by 2.0 seconds until the height is less than 50 feet; then increment time by 0.05 seconds. Stop the program if the rocket impacts or if the total time exceeds 100 seconds.

 2. Input/Output Description

Input—none
Output—a table of time and height values

 3. Hand Example

Using a calculator, compute the first three entries in the table as shown:

Time (s)	Height (ft)
0.0000	60.0000
2.0000	68.5001
4.0000	93.7719

We are now ready to develop the algorithm so that the computer can compute the rest of the table for us.

 4. Algorithm Development

Decomposition

Set time to zero.
Compute and print times and heights.

We do the refinement for this problem in two steps.

Initial Pseudocode

rocket: time ← 0
            while above ground and time ≤ 100 do
                    compute height
                    print time, height
                    increment time

In our refinement we replace "increment time" with the steps that take into account our height above ground. We also replace the condition "above ground" with a specific condition based on our height above ground.

Final Pseudocode

```
rocket:    time ← 0
           height ← 60
           while height > 0 and time ≤ 100 do
                   compute height
                   print time, height
                   if height < 50 then
                           add 0.05 to time
                   else
                           add 2.0 to time
```

Notice that we initialized the height variable to 60.0 before entering the while loop. Why?

Fortran Program

```fortran
!-----------------------------------------------              ---------!
program rocket
!
!   This program simulates a rocket flight for up to
!   100 seconds. A table of data is generated containing
!   height values at increments of 2 seconds until the
!   rocket is within 50 feet of the ground. Within 50
!   feet of the ground, the height values are computed
!   every 0.05 second.
!
!
!    Declare and initialize variables.
     implicit none
     real :: height=60.0, time=0.0
!
!    Print report headings.
     print 5
   5 format(1x,"Time (s)    Height (ft)"/)
!
!    Generate and print report information.
     do while ((height > 0.0).and.(time <= 100.0))
        height = 60.0 + 2.13*time**2 - 0.0013*time**4 &
                + 0.000034*time**4.751
        print 15, time, height
  15    format (1x,f7.4,6x,f9.4)
        if (height < 50.0) then
           time = time + 0.05
        else
           time = time + 2.0
        end if
     end do
!
end program rocket
!----------------------------------------------------------------------!
```

5. Testing

The first few lines and the last few lines of output are shown. The values are in agreement with the hand example.

```
Time (x)    Height (ft)

 0.0000       60.0000
 2.0000       68.5001
 4.0000       93.7719
 6.0000      135.1644
 8.0000      191.6590
 . . .
54.0000      999.1536
56.0000      827.4205
58.0000      633.2993
60.0000      418.3975
62.0000      184.8088
64.0000      -64.8608
```

Can you think of ways to test different parts of the algorithm? We now know that the rocket impacts before 100 seconds of flight time; we could change the cutoff time to 50 seconds to see if this exit from the while loop were working correctly. How could you modify the program to check the change in the increment of the time variable from 2.0 seconds to 0.05 second?

3-6 DO LOOP

In Section 3-4 we used the do while statement to build while loops. A special form of the while loop is the counting loop, or *iterative loop*. Implementing a counting loop generally involves initializing a counter before entering the loop, modifying the counter within the loop, and exiting the loop when the counter reaches a specified value. Counting loops are executed a specified number of times. The three steps (initialize, modify, and test) can be incorporated in a while loop, but they still require three different statements. The do statement, in contrast, combines all three steps into one. Using the do statement to construct a loop results in a construction called a *do loop.*

The general form of the do statement is

```
do index = initial, limit, increment
      statement 1
      ...
      statement n
end do
```

The *index* is a variable used as the loop counter; *initial* represents the initial value given to the loop counter; *limit* represents the value used to determine when the do loop has been completed; and *increment* represents the value to be added to the loop counter each time the loop is executed.

The initial, limit, and increment values are the *parameters* of the do loop. If the increment is omitted, an increment of 1 is assumed. When the value of the index is greater than the *limit*, control is passed to the statement following the end of the loop (assuming that the increment is a positive value). Before listing the rules for using a do loop, let's look at a simple example.

EXAMPLE 3-4

Integer Sum

The sum of the integers 1 through 50 is represented mathematically as

$$\sum_{i=1}^{50} i = 1 + 2 + \cdots + 49 + 50$$

Obviously, we do not want to write one long assignment statement to compute this sum. A better solution is to build a loop that executes 50 times and adds a number to the sum each time, as shown in the following solutions.

SOLUTION

While Loop Solution
```
integer :: number=1, sum=0
. . .
do while (number <= 50)
    sum = sum + number
    number = number + 1
end do
```

Do Loop Solution
```
integer :: number, sum=0
. . .
do number=1,50
    sum = sum + number
end do
```

The do statement initializes number to 1. The loop is repeated until the value of number is greater than 50. Because the third parameter is omitted, the index number is incremented automatically by 1 at the end of each loop. Comparing the do loop solution with the while loop solution, we see that the do loop solution is shorter but that both compute the same value for sum.

. . . .

Structure of a Do Loop

We have seen a do loop in a simple example. The following list summarizes the rules related to its structure.

1. The index of the do loop must be a variable, but it may be either real or integer.
2. The parameters of the do loop may be constants, variables, or expressions and can also be real or integer types.
3. The increment can be either positive or negative, but it cannot be zero.

In the following pseudocode we show a counting loop using a while loop structure:

$$\text{index} \leftarrow \text{initial value}$$
$$\text{while (index} \leq \text{limit)}$$
$$\text{statements}$$
$$\text{add increment to index}$$

The flowchart symbol for a do loop is

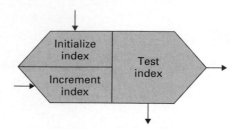

Note that the symbol is divided into three parts, each corresponding to the three steps in building a counting loop. The flowchart for the counting loop in Example 3-4 is

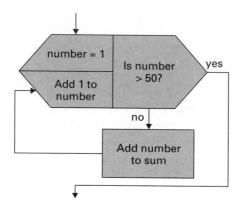

Execution of a Do Loop

In the following list we present a complete set of rules related to do loop execution. We suggest that you read through them once and then go through the set of examples that follow the rules. Later, reread these rules. Do not worry about memorizing them; after you have gone through the set of examples, most of the rules will seem logical.

1. The test for completion is done at the beginning of the loop, as in a while loop. If the initial value of the index is greater than the limit and the increment is positive, the loop will not be executed. For instance, the following statements specify an initial value of 5 for the index i. This value is greater than the limit, 2; therefore, the statements within the loop will be skipped and control passes to the statement following this do loop.

```
integer :: i
...
do i=5,2
    statements
end do
```

2. The value of the index should not be modified by other statements during the execution of the loop.
3. After the loop begins execution, changing the values of variables used as initial, limit, or increment parameters will have no effect on the loop.
4. If the increment is negative, the exit from the loop will occur when the value of the index is less than the limit.
5. If a cycle statement is executed within a do loop, control will pass back to the beginning of the loop. The index will be incremented, and its value will be compared to the limit to determine if the loop should be executed again.
6. If an exit statement is executed within a do loop, control will pass to the statement following the do loop. The index contains the value that it had when the exit statement was executed.
7. Upon completion of the do loop, the index contains the last value that exceeded the limit.
8. The number of times that a do loop will be executed can be computed as

$$\left[\frac{\text{limit} - \text{initial}}{\text{increment}}\right] + 1$$

The brackets around the fraction represent the greatest integer value; that is, we drop (truncate) any fractional portion of the quotient. If this value is negative, the loop is not executed. If we had the following do statement

```
do k=5,83,4
   . . .
end do
```

the corresponding do loop would be executed the following number of times:

$$\left[\frac{83 - 5}{4}\right] + 1 = \left[\frac{78}{4}\right] + 1 = 19 + 1 = 20$$

The value of the index k would be 5, then 9, then 13, and so on until the final value of 81. The loop would not be executed with the value 85 because it is greater than the limit, 83.

The next example illustrates both the structure and the execution of the do loop.

EXAMPLE 3-5 **Polynomial Model**

Polynomials are often used to model data and experimental results. Assume that the polynomial $3t^2 + 4.5$ models the results of an experiment, where t represents time in seconds. Write a program to evaluate this polynomial for time beginning at zero seconds and ending at 5 seconds in increments of 0.5 second.

SOLUTION

Step 1 is to state the problem clearly:

> Print a report to evaluate the polynomial $3t^2 + 4.5$ for a period from 0 seconds through 5 seconds in increments of 0.5 second.

Step 2 is to describe the input and output:

Input — none

Output — report containing polynomial values

Step 3 is to work a simple example. Therefore, the first few lines of output should be

```
              Polynomial Model


              Time   Polynomial
              (s)
              0.0      4.50
              0.5      5.25
              1.0      7.50
```

Step 4 is to develop an algorithm, beginning with the decomposition.

Decomposition

```
┌─────────────────────┐
│ Print headings.     │
├─────────────────────┤
│ Print report.       │
└─────────────────────┘
```

Pseudocode

```
polynomial_model:    print headings
                     k ← 0
                     while k ≤ 50
                         time ← 0.1 · k
                         polynomial_1 = 3 · time² + 4.5
                         print time, polynomial_1
                         add 5 to k
```

Fortran Program

```fortran
!----------------------------------------------------------------------!
program polynomial_model
!
!  This program prints a table of values for a
!  polynomial starting with time equal to 0
!  seconds through 5 seconds, in increments
!  of 0.5 second.
!
!
!    Declare and initialize variables.
     implicit none
     integer :: k
     real :: polynomial_1, time
!
!    Print report heading.
```

```
      print*, "Polynomial Model"
      print*
      print*, "Time    Polynomial"
      print*, "(s)"
!
!     Compute and print polynomial values.
      do k=0,50,5
         time = 0.1*real(k)
         polynomial_1 = 3.0*time**2 + 4.5
         print 10, time, polynomial_1
  10     format(1x,f4.1,4x,f5.2)
       end do
!
end program polynomial_model
!-----------------------------------------------------------------!
```

Note that the value of time is computed from an integer that varies from 0 to 50 in steps of 5. Another possibility would have been to use the following do statement:

$$do\ time=0.0,5.0,0.5$$

However, it is wise to avoid do statements such as this one with real parameters. These statements do not always execute exactly the way we expect because of truncation within the computer. For example, suppose that the value for 0.5 is stored as a value slightly less than 0.5 in our computer system. Each time we add 0.5 to the index, we are adding less than we intend, and the values are not those we intend.

Step 5 is to test the program. The output from this program is

```
Polynomial Model

Time   Polynomial
(s)
0.0       4.50
0.5       5.25
1.0       7.50
1.5      11.25
2.0      16.50
2.5      23.25
3.0      31.50
3.5      41.25
4.0      52.50
4.5      65.25
5.0      79.50
```

. . . .

Try It

Try this self-test to check your memory of some key points from Section 3-6. If you have any problems with the exercises, you should reread this section. The solutions are given at the end of this module.

In problems 1–6 determine the number of times that the statements in the do loop will be executed. Assume that the index is an integer variable.

1. do number=5,14

2. do count=-4,4

3. do k=15,3,-1

4. do time=-5,15,3

5. do time=50,250,25

6. do index=72,432,4

For problems 7–11 give the value in count after each of the following loops is executed. Assume that count is an integer variable initialized to zero before each problem.

7. do i=1,8
        count = count + 1
    end do

8. do k=1,5
        count = count + k
    end do

9. do index=0,7
        count = count - 2
    end do

10. do number=8,0,-1
        count = count + 2
    end do

11. do m=-5,5
        count = count + (-1)**m
    end do

3-7 Application　　**TIMBER REGROWTH**

Environmental Engineering

A problem in timber management is to determine how much of an area to leave uncut so that the harvested area is reforested in a certain period of time. It is assumed that reforestation takes place at a known rate per year, depending on climate and soil conditions. A reforestation equation expresses this growth as a function of the amount of timber standing and the reforestation rate. For example, if 100 acres are left standing after harvesting and the reforestation rate is 0.05, then $100 + 0.05 \times 100$, or 105 acres, are forested at the end of the first year. At the end of the second year, the number of acres forested is $105 + 0.05 \times 105$, or 110.25 acres.

Write a program to read the identification number of an area, the total number of acres in the area, the number of acres that are uncut, and the reforestation rate. Print a report that tabulates for 20 years the number of acres reforested and the total number of acres forested at the end of each year.

1. Problem Statement

Compute the number of acres forested at the end of each year for 20 years for a given area.

2. Input/Output Description

Input — the identification number for the area of land, the total acres, the number of acres with trees, and the reforestation rate.

Output—a table with a row of data for each of 20 years. Each row of information contains the number of acres reforested during that year and the total number of acres forested at the end of the year.

3. Hand Example

Assume that there are 14,000 acres total with 2500 acres uncut. If the reforestation rate is 0.02, we can compute a few entries as shown:

Year 1 $2500 \times 0.02 = 50$ acres of new growth
original 2500 acres + 50 new acres = 2550 acres forested

Year 2 $2550 \times 0.02 = 51$ acres of new growth
original 2550 acres + 51 new acres = 2601 acres forested

Year 3 $2601 \times 0.02 = 52.02$ acres of new growth
original 2601 acres + 52.02 new acres = 2653.02 acres forested

4. Algorithm Development

The overall structure is a counting loop that is executed 20 times, once for each year. Inside the loop we need to compute the number of acres reforested during that year and add that number to the acres forested at the beginning of the year; this will compute the total number of acres forested at the end of the year. The output statement should be inside the loop, because we want to print the number of acres forested at the end of each year.

Decomposition

Read initial information.
Print headings.
Print report.

Initial Pseudocode

timber_management: read initial information
                       print headings
                       year ← 1
                       while year ≤ 20
                            compute reforested amount
                            add reforested amount to uncut amount
                            print reforested amount, uncut amount
                            add 1 to year

Clearly, an error condition exists if the uncut area exceeds the total area. We will test for this condition and exit after printing an error message if it occurs. It is possible to imagine soil conditions that would result in a zero or negative reforestation rate, so we will not perform any error

checking on the rate. However, all the values read will be printed, or
echoed, so that the user can recognize an error in an input value. We now
add these refinements to our initial pseudocode.

Final Pseudocode

timber_management: read identification, total, uncut, rate
                      print identification, total, uncut, rate
                      if uncut > total then
                              print error message
                      else
                              print headings
                              year ← 1
                              while year ≤ 20
                                      reforested ← uncut · rate
                                      add reforested to uncut
                                      print year, reforested, uncut
                                      add 1 to year

Fortran Program

```fortran
!---------------------------------------------------------------!
program timber_management
!
!  This program computes a reforestation summary for an
!  area that has not been completely harvested.
!
!
!     Declare and initialize variables.
      implicit none
      integer :: id, year
      real :: rate, reforested, total, uncut
!
!     Prompt user for information.
      print*, "Enter land identification (integer):"
      read*, id
      print*, "Enter total number of acres:"
      read*, total
      print*, "Enter number of acres uncut:"
      read*, uncut
      print*, "Enter reforestation rate:"
      read*, rate
!
!     Generate and print reforestation information.
      if (uncut > total) then
         print*, "Uncut area larger than entire area."
      else
         print 5, id, total, uncut, rate
    5    format(/1x,"Reforestation Summary "/ &
               1x,"Identification Number ",i5/ &
               1x,"Total Acres ",f10.2/ &
               1x,"Uncut Acres ",f10.2/ &
               1x,"Reforestation Rate ",f5.3// &
               1x,"Year  Reforested  Total Reforested")
```

```
          do year=1,20
             reforested = uncut*rate
             uncut = uncut + reforested
             print 10, year, reforested, uncut
10           format(1x,i3,f11.3,f17.3)
          end do
       end if
!
end program timber__management
!-----------------------------------------------------------------!
```

---

5. Testing

Using the test data from the hand example, a typical interaction is

```
Enter land identification (integer):
25563
Enter total number of acres:
14000.0
Enter number of acres uncut:
2500.0
Enter Reforestation rate
0.02

Reforestation Summary
Identification Number 25563
Total Acres     14000.00
Uncut Acres      2500.00
Reforestation Rate 0.020

Year  Reforested   Total Reforested
  1      50.000         2550.000
  2      51.000         2601.000
  3      52.020         2653.020
  4      53.060         2706.080
  5      54.122         2760.202
  6      55.204         2815.406
  7      56.308         2871.714
  8      57.434         2929.148
  9      58.583         2987.731
 10      59.755         3047.486
 11      60.950         3108.436
 12      62.169         3170.604
 13      63.412         3234.017
 14      64.680         3298.697
 15      65.974         3364.671
 16      67.293         3431.964
 17      68.639         3500.604
 18      70.012         3570.615
 19      71.412         3642.028
 20      72.841         3714.868
```

The numbers match the ones we computed by hand. Try an example to test the error condition by using an uncut area larger than the total area. What happens if the reforestation rate is 0.00 or − 0.02? Should there be an upper limit on the reforestation rate? This information is not given in the original problem, so we probably should not set one arbitrarily. What happens if you enter 14,000.0 instead of 14000.0? Try it. It might be a good idea to remind the program user not to use commas in numbers. How would you do this?

3-8 NESTED DO LOOPS

Do loops can be nested within other do loops, just as we used if structures within other if structures. The following rules apply to writing and executing *nested do loops*:

1. A nested do loop cannot use the same index as a loop that contains it.
2. A nested do loop must be completely within the outer do loop. For example, if a do statement is within another do loop, the end do statement for the nested loop must also be within the outer loop.
3. Do loops that are independent of each other may use the same index, even if they are all contained within another do loop.
4. When one loop is nested within another, the inside loop is completely executed with each pass through the outer loop.

The following are a set of valid loops and a set of invalid loops:

Valid Loops

```
do i=1,5                    do i=1,5
   do j=1,8                     do k=1,8
      do k=2,10,2                  . . .
         . . .                  end do
      end do                  do k=2,10,2
   end do                        . . .
end do                        end do
                           end do
```

Invalid Loops (same index for dependent loops)

```
do i=1,5                    do j=1,5
   do j=1,8                     do j=1,8
      do i=2,10,2                  . . .
         . . .                  end do
      end do                  end do
   end do
end do
```

To illustrate the execution of a program with nested loops, consider the following:

```
!------------------------------------------------------------------!
program nest
!
!  This program prints the indices in nested do loops.
!
!
!    Declare and initialize variables.
     implicit none
     integer :: i, j
!
!    Print indices inside nested loops.
     print*, "i    j"
     print*
     do i=1,5
        do j=3,1,-1
           print 5, i, j
   5       format(1x,i1,3x,i1)
        end do
        print*, "end of pass"
     end do
!
end program nest
!------------------------------------------------------------------!
```

The output is the following:

```
i   j

1   3
1   2
1   1
end of pass
2   3
2   2
2   1
end of pass
3   3
3   2
3   1
end of pass
4   3
4   2
4   1
end of pass
5   3
5   2
5   1
end of pass
```

The first time through the outer loop, i is initialized to the value 1. We thus begin executing the inner loop: The variable j is initialized to the value 3. After executing the print 5 statement, we reach the end of the inner loop, and j is decremented by 1 to the value 2. Because 2 is still larger than the final value of 1, we repeat the loop. The variable j is decremented by 1 to the value 1, and we repeat the loop again. When j is decremented again, it is less than the final value of 1, so we have completed the inner loop and the message end of pass is printed. The variable i is incremented to 2, and we begin the inner loop again. This process is repeated until i is greater than 5.

EXAMPLE 3-6   ## Experimental Sums

Write a complete program to read 20 data values. Compute the sum of the first 5 values, the next 5 values, and so on. Print the 4 sums. Assume that the values are real.

SOLUTION

Step 1 is to state the problem clearly:

> Read 20 data values and compute the sum of the first 5 values, the second 5 values, and so on.

Step 2 is to describe the input and output:

Input — 20 data values
Output — sum of the first 5 values, sum of the second set of 5 values, and so on for all 20 data values

Step 3 is to work a simple example by hand. Assume we have 6 values and we want to sum them in groups of 2. Then we have 3 sums to compute, and for each sum we add 2 values.

$$\text{values:} \quad 4, 6, 1, 2, 7, -2$$

The corresponding sums are 10, 3, and 5.

Step 4 is to develop an algorithm. Although we will read a total of 20 data values, we need only 5 values at a time; thus, we need an outer loop to read four sets of data. Each set of data is 5 values; thus, the inner loop reads the 5 values. It is important to initialize to zero the variable being used to store the sum before the inner loop is begun. We then add 5 values, print the sum, and set the sum back to zero before we read the next set of values.

The decomposition is a single step in this problem; all the other steps are performed inside the overall loop.

Decomposition

> Read data, and compute and print sums.

The refinement in pseudocode illustrates the structure of this algorithm, which is a loop within a loop.

Pseudocode

```
sums:    i ← 1
         while i ≤ 4
             j ← 1
             while j ≤ 5
                 read value
                 add value to sum
                 add 1 to j
             print sum
             add 1 to i
```

We can now translate the pseudocode into Fortran.

Fortran Program

```fortran
!-----------------------------------------------------------------!
program sums
!
!   This program reads 20 values and prints
!   the sum of each group of 5 values.
!
!
!     Declare and initialize variables.
      implicit none
      integer :: i, j
      real :: sum, value
!
!     Nested loops.
      do i=1,4
         sum = 0.0
!
         do j=1,5
            print*, "Enter data value:"
            read*, value
            sum = sum + value
         end do
!
         print 10, i, sum
 10      format(1x,"sum ",i1," = ",f6.2)
      end do
!
end program sums
!-----------------------------------------------------------------!
```

Step 5 is to test the program. Here are the first few lines of output from this program:

```
Enter data value:
10
Enter data value:
2.40
Enter data value:
8.0
Enter data value:
3.0
Enter data value:
0.04
sum 1 =   23.44
Enter data value:
...
```

. . . .

EXAMPLE 3-7

Factorial Computation

Write a complete program to compute the factorial of an integer. A few factorials and their corresponding values are shown (an exclamation point following a number symbolizes a factorial):

$0! = 1$

$1! = 1$

$2! = 2 \times 1$

$3! = 3 \times 2 \times 1$

Compute and print the factorial for four values read from the terminal.

SOLUTION

Step 1 is to state the problem clearly:

> Compute the factorial of values read from the terminal.

Step 2 is to describe the input and output:

Input—four data values
Output—factorial of each of the four data values

Step 3 is to work a simple example by hand:

$$5! = 5 \times 4 \times 3 \times 2 \times 1 = 120$$

Step 4 is to develop an algorithm. The factorial of a negative number is not defined, so we should include an error check in the algorithm for this condition. In computing a factorial, we use a counting loop to perform the successive multiplications. The overall structure of this problem solution is a loop, so the decomposition is again a single step.

Decomposition

Read values, and compute and print factorials.

Pseudocode

factorial: i ← 1
              while i ≤ 4
                  read n
                  if n < 0
                      print error message
                  else
                      n_factorial ← 1
                      if n > 1 then
                          k ← 1
                          while k ≤ n
                              multiply n_factorial by k
                              add 1 to k
                      print n, n_factorial
                  add 1 to i

Fortran Program

```
!----------------------------------------------------------------!
program factorial
!
!   This program computes the factorial
!   of four values read from the terminal.
!
!
!     Declare variables.
      implicit none
      integer :: i, k, n, n_factorial
!
!     Read four values, and compute the
!     corresponding factorials.
      do i=1,4
          print*, "Enter n:"
          read*, n
!
          if (n < 0) then
!
              print 5, n
    5         format(1x,"Invalid n = ",i7)
!
          else
!
              n_factorial = 1
              if (n > 1) then
                  do k=1,n
                      n_factorial = n_factorial*k
                  end do
              end if
              print 10, n, n_factorial
   10         format(1x,i4,"! = ",i8)
!
          end if
      end do
!
end program factorial
!----------------------------------------------------------------!
```

Step 5 is to test the program. The output from a sample run is

```
Enter n:
3
    3! =          6
Enter n:
-2
Invalid n =        -2
Enter n:
11
   11! = 39916800
Enter n:
0
    0! =          1
```

The value of a factorial of a large integer can quickly exceed the limits of integer variables. Therefore, to compute the factorial of a wide range of integers correctly, use real variables for the computations.

. . . .

Try It

Try this self-test to check your memory of some key points from Section 3-8. If you have any problems with the exercises, you should reread this section. The solutions are included at the end of this module.

For problems 1 – 3 give the value in count after each of the following loops is executed. Assume that count is initialized to zero before starting each problem. Also assume that the index of each loop is an integer.

1. ```
do in=5,15
    do k=2,0,-1
        count = count + 1
    end do
end do
```

2. ```
do k=0,5
    do j=5,-5,-2
        count = count + 1
    end do
end do
```

3. ```
do i=5,14,2
    do k=4,0
        count = count + 1
    end do
    count = count + 1
end do
```

**SUMMARY**

This chapter greatly expanded the types of problems we can solve in Fortran, because if statements can control the order in which statements are executed. We also learned to use the do statement to implement both while loops and counting loops. Most of the programs in the rest of the module will use one of these loop structures.
```

Key Words

compound logical expression	logical operator
condition	logical variable
control structure	loop
counting loop	nested do loop
do loop	parameter
flowchart	pseudocode
hierarchy	relational operator
if structure	repetition
increment value	selection
index	sequence
initial value	structured algorithm
iterative loop	structured program
limit value	while loop
logical expression	

Problems

This problem set begins with modifications to programs given earlier in this chapter. Give the decomposition, refined pseudocode or flowchart, and Fortran program for each problem.

Problem 1 modifies the program light_pipe given in Section 3-3 on page 68.

1. Modify the light pipe program so that it reads the refraction index for the rod material. Then, for an incidence angle of 45°, give the range of refraction indices for materials that will create light pipes.

Problem 2 modifies the trajectory program rocket given in Section 3-5 on page 75.

2. Modify the rocket trajectory program so that it reads two real values, increment_1 and increment_2. Start time at zero seconds and increment it by increment_1 seconds until the distance is less than 50 feet; then increment time by increment_2 seconds.

Problems 3-4 modify the regrowth program timber_management, given in Section 3-7 on page 84.

3. Modify the timber regrowth program so that an integer value M is read from the terminal, where M represents the number of years that should be between lines in the output table.
4. Modify the timber regrowth program so that instead of printing data for 20 years, it determines the number of years required for the cut area to be completely reforested.

For problems 5-7 write complete Fortran programs to print tables showing the values of the input variables and the function shown using do loops to control the loops.

5. Print a table of values for the real variable k where

$$k = |\ j - 3j + \sqrt{j}\ |$$

for values of $j = 0, 1, 2, \ldots, n$; n is read from the terminal. (Assume j and n are integers.)

6. Print a table of values for y where

$$y = \cos x$$

for values of $x = 0.0, 10.0, \ldots, 350.0$; the units of x are in degrees. (Assume x and y are real variables.)

7. Print a table of values for f where

$$f = xy - 1$$

for values of $x = 1, 2, \ldots, 9$, with corresponding values of $y = 0.5, 0.75, 1.0, \ldots, 2.5$. (Assume f, x, and y are real values.)

Develop new programs in problems 8–10. Use the five-step problem-solving process.

8. Write a program to read three integers, i, j, and k. Then determine if the integers are in ascending order ($i \leq j \leq k$), descending order ($i \geq j \geq k$), or neither order. Print an appropriate message.

9. A set of 20 temperatures has been recorded during an experiment. We want to compute and print the average temperature, and we also want to determine if a new record low or record high temperature occurred during this particular experiment. Therefore, on one line we enter the previous low temperature and the previous high temperature. The remaining 20 temperatures are entered one per line. Read this information and print the average. If a new record low or high occurred, print an appropriate message.

10. Consider the trajectory of a stone hurled into the air with an initial velocity v at an angle to the horizontal of θ. Neglecting drag due to friction with the atmosphere, the following equations describe the stone's distance (d) from the initial spot and the height (h) of the stone at time t:

$$d = v \cdot t \cdot \cos \theta$$
$$h = v \cdot t \cdot \sin \theta - 0.5 \cdot g \cdot t^2$$

where g is the acceleration due to gravity (9.8 m/s² or 32 ft/s²). Write a program to read the initial velocity and angle, and then print a table of distance and height values. Use intervals of 0.25 second between values, and stop when the height becomes negative.

4 Engineering and Scientific Data Files

Critical Path Analysis Critical path analysis is a technique used to determine the time schedule for large projects, ranging from the construction of a large building to the construction of the new 777 jumbo jet. This information is important in the planning stages before a project is begun, and it is also useful to evaluate the progress of a project that is partially completed. One method for this analysis starts by breaking a project into sequential events and then breaking each event into various tasks. Although one event must be completed before the next one is started, various tasks within an event can occur simultaneously. The time it takes to complete an event, therefore, depends on the number of days required to finish its longest task. Similarly, the total time it takes to finish a project is the sum of the time it takes to finish the individual events.

INTRODUCTION

ngineers and scientists must frequently analyze large amounts of data. It is not reasonable to enter all this data into the computer by hand each time it is needed. We can enter the data once into a *data file;* then each time the data is needed, we can read it from the data file. In addition to printing data for a report, we can write it into a data file; other programs can then easily access the data. This chapter focuses on reading information from existing data files.

4-1 I/O STATEMENTS

In the programs in previous chapters, we used `read` statements when we needed information entered during the execution of the program. For small amounts of data, this interaction through the keyboard is satisfactory. However, for large amounts of data, it is not feasible to enter the data every time we need to run the program. A data file is a file that contains data that can be read by a program. A data file can be built in two ways: It can be generated by another program, or it can be generated using a word processor or an editor. Each line of the data file is a *record.* A record in a Fortran program file contains a Fortran language instruction, while a record in a data file contains data values. Once a file is generated, it can be listed or updated using an editor or a word processor. For now, our data files will contain only numbers; Chapter 7 covers how to work with character information.

To use data files with our programs, we must use some new statements and some extensions of old statements. Each of these statements references the filename that is assigned when a file is built. If we build a data file with a word processor or an editor, we assign the filename when we enter the data. If we build a data file with a program, we include a statement in the program that gives the file a name.

If a file is going to be used in a program, it must be opened before any processing can be done with the file. The `open` statement gives the program several important pieces of information about the file, including the name of the file and whether the file already exists. (Restrictions on filenames are system dependent. We will use a maximum of eight characters plus a file extension, as in `results.dat`.) In addition, the `open` statement connects the file to the program using a unit number that corresponds to the file. Then, within the program, whenever we want to refer to the file, we use the unit number. Since there may be several files used in the same program, the unit number gives us a method for uniquely specifying a file. The general form of the `open` statement that we use is

open(unit=*integer expression,* file=*filename,* status=*literal*)

The integer expression is the unit number assigned to the file, the filename is the name given to the file when it was generated, and the literal tells the computer whether the file already exists. If the file already exists, it is specified with `status="old"`; if the file does not exist, it is specified with `status="new"`; if the existence of the file is not known, it is specified with `status="unknown"`.

The open statement should precede any read or write statements that use the file. Also, in general, the open statement should be executed only once; therefore, it should not be inside a loop. Some systems require a rewind statement after opening an input file to position the file at its beginning.

To read from a data file, we use an extension of the list-directed read statement that has the following general form:

> read*(unit number,*)* variable list

To write information to a data file, we use a new statement—a write statement. Just like the print statement, the write statement can be used with list-directed output or formatted output. The list-directed write statement has the following general form:

> write*(unit number,*) expression list*

The formatted write statement has the following general form:

> write*(unit number, k) expression list*

where *k* is the statement number of the corresponding format statement. In all of these general forms, the unit number corresponds to the unit number assigned in the open statement. The asterisk following the unit number specifies that we are using list-directed input or output.

Most computer systems have several input or output devices attached to them. Each device is assigned a unit number. For example, if a laser printer has been assigned unit number 8, then the following statement would write the values of x and y using the laser printer:

```
write(8,*) x, y
```

Many systems assign the standard input device (typically the terminal keyboard) to unit number 5 and the standard output device (typically the terminal screen) to unit number 6; these devices are used when your program executes read* or print* statements. Avoid using preassigned unit numbers with your data files; if your computer system assigns unit numbers 5 and 6 as previously defined, do not use them with your data files. Otherwise, errors could result if your program assigned a data file to unit 6 and the program then executed a print* statement. The name of the file is determined when the file is built, but we may choose any unit number to refer to the file as long as the number is not preassigned by our computer system. Thus, in one program we could use unit 10 to refer to a file of experimental velocity measurements, and then in another program we could use unit 12 to refer to the same file.

When we finish executing a program that has used files, the files are automatically closed before the program terminates. Occasionally you may want to specifically close, or disconnect, a file from your program. The Fortran statement to close a file has the following general form:

```
close(unit=integer expression)
```

The following list summarizes some important rules to remember when reading data from data files:

1. Each read statement starts with a new line of data. If any values left on the previous line were not read, they are skipped.
2. If a line does not contain enough values to match the list of variables on the read statement, another data line is automatically read to acquire more values. Additional data lines are read until values have been acquired for all the variables listed on the read statement.
3. A read statement does not have to read all the values on the data line; however, it does have to read all the values on the same line that are before the values that you want. For example, if a file has five values per line and you are interested in the third and fourth values, you must read the first and second values to get to the third and fourth values, but you do not need to read the fifth value.

To use the correct read statement, you must know how the values were entered in the data file. For example, assume that a data file contains two numbers per line, representing a time and a temperature measurement. Also assume that the first three lines of the file contain the following information:

line 1:   0.0   89.5
line 2:   0.1   90.3
line 3:   0.2   90.8

The following statement correctly reads a time value and temperature value from the data file:

```
read(10,*) time, temperature
```

It would be incorrect to use the following two statements in place of the previous statement:

```
read(10,*) time
read(10,*) temperature
```

The execution of these two statements reads two lines from the data file, not one. For example, if the two statements are the first read statements to be executed using this file, the value of time will be 0.0 and the value of temperature will be 0.1. Not only will we have the wrong values in the variables, but there will not be an error message to let us know that something is wrong. This example illustrates the importance of carefully testing our programs with known data before using them with other data files.

Try It   Try this self-test to check your memory of some key points from Section 4-1. If you have any problems with the exercises, you should reread this section. The solutions are given at the end of this module.

The data file `lab1.dat` contains the following information, which represents a time and temperature for each line:

line 1:   0.0   86.3
line 2:   0.5   93.5
line 3:   1.0   95.5
line 4:   1.5   97.8
line 5:   2.0   98.2
line 6:   2.5   99.1

Give the values of the variables after executing the following statements. For each problem assume that the data file has been opened, but no previous read statements have been executed. Assume all variables have been defined as real variables.

1. `read(10,*) time, temperature`
2. `read(10,*) time`
   `read(10,*) temperature`
3. `read(10,*) time_1, temperature_1, time_2, temperature_2`
4. `read(10,*) time_1, temperature_1`
   `read(10,*) time_2, temperature_2`
5. `read(10,*) time_1`
   `read(10,*) temperature_1`
   `read(10,*) time_2`
   `read(10,*) temperature_2`
6. `read(10,*) time_1, time_2`
   `read(10,*) temperature_1, temperature_2`

4-2 TECHNIQUES FOR READING DATA FILES

To read information from a data file, we must first know some information about the file. In addition to the name of the file, we must know what information is stored in the file. This information must be very specific, such as the number of values entered per line and the units of the values. In addition, we need to know if any special information in the file will be useful in deciding how many records are in the file or how to determine when we have read the last record. This information is important because if we execute a read statement after we have already read the last record in the file, we will get an execution error and the program will quit. We can avoid an execution error by using the information that we have about the file to decide what kind of loop we should use in the program. For example, if we know that there are 200 records in the file, then we can use a do loop that is executed 200 times, and each time through the loop we read a record and perform the desired computations with the information. Often, we do not know ahead of time how many records are in the data file, but we know that the last record contains special values so that our program can test for them. For example, if a file contained time and temperature measurements, the last record could contain values of −999 for both measurements to signal that it is the last record. Then we could exit a while loop when the time and temperature values are −999. Finally, a third situation arises when we do not know ahead of time how many records are in a file, and no special value appears at the end of the file. A special option in the read statement will be discussed that allows us to handle this situation.

We now look at each of these three cases separately. For each case we use the same example so that we can compare the solutions.

Specified Number of Records

The first case we consider is one in which we know or we can ascertain the number of records in the data file. If we know the exact number of records, a do loop can be used to process the data. Sometimes we generate files with special information, such as the number of valid data records, in the first record in the data file. To process this file, we read the number at the beginning of the file using a variable such as count. Then we can execute a do loop with count as part of the do statement. The following example illustrates reading a data file with a record count in the first record.

EXAMPLE 4-1

Average of Laboratory Data

Assume that we have a data file named results1.dat that contains information collected from a laboratory experiment. Each line of the data file, except the first line, contains two numbers: a time value in seconds and a temperature measurement in degrees Fahrenheit. The first line contains an integer that represents the number of time and temperature pairs that follow in the data file. Read the data and compute the average temperature value. Print the number of data values and the average value.

SOLUTION

Step 1 is to state the problem clearly:

Compute and print the average of a set of temperature measurements.

Step 2 is to describe the input and output:

Input—time and temperature values in a data file named results1.dat
Output—the average of the temperature values and a count of the temperature values

Step 3 is to work a simple example by hand. Assume that the data file contains the following information:

```
4
0.0        120
0.5        132
1.0        144
1.5        163
```

For this data, the total number of values is 4, and the average temperature is (120+132+144+163)/4, or 139.75 degrees Fahrenheit.

Step 4 is to develop an algorithm, starting with the decomposition.

Decomposition

Read and compute a sum of temperature values.
Compute the average temperature.
Print the count and the average temperature.

Pseudocode

```
solution_1:      sum ← 0
                 read count
                 if count < 1 then
                         print error message
                 else
                         j ← 1
                         while j ≤ count
                                 read time, temperature
                                 add temperature to sum
                                 add 1 to j
                         average ← sum/count
                         print count, average
```

Fortran Program

```fortran
!--------------------------------------------------------------!
program solution_1
!
!  This program computes the average temperature
!  using a set of times and temperatures stored in
!  a data file. The first line contains the number
!  of actual data records that follow.
!
!
!     Declare and initialize variables.
      implicit none
      integer :: count, j
      real :: sum=0.0, temperature, time
!
!     Open file and read count.
      open(unit=10,file="results1.dat",status="old")
      read(10,*) count
!
!     Check for invalid count.
```

```
       if (count < 1) then
          print*, "No data according to record count"
       else
!
!         Read data values and compute average.
          do j=1,count
             read(10,*) time, temperature
             sum = sum + temperature
          end do
          print 25, count, sum/real(count)
   25     format(1x,"count = ",i3,5x,"average = ",&
                 f8.2," degrees F")
!
       end if
!
!      Close file.
       close(unit=10)
!
end program solution_1
!-------------------------------------------------------------------!
```

Step 5 is to test the program. If we test this program using the hand example, the following information is printed:

```
count =    4      average =    139.75 degrees F
```

. . . .

Trailer or Sentinel Signals

Special data values that are used to signal the last record of a file are called *trailer*, or *sentinel*, *signals*. When we generate the file, we add the special data values in the last record. When using a file with a trailer signal, we need to be careful that our program does not treat the information on the trailer line as regular data. For example, if we are counting the number of valid lines of data in the file, we want to be sure that we do not count the trailer line. Also, it is important to include the same number of data values on the trailer line as are on the rest of the lines in the data file. Since the trailer line is read with the statement that reads the other lines, it will try to read the line after the trailer line if there are not enough values on the trailer line; this will cause an execution error.

The following example illustrates reading a data file with a trailer record.

EXAMPLE 4-2 ## Average of Laboratory Data

Assume that we have a data file named `results2.dat` that contains information collected from a laboratory experiment. Each line of the data file contains two numbers: a time value in seconds and a temperature measurement in degrees Fahrenheit. However, the last line of the file is a trailer signal that contains − 999 for the time and temperature values. Read the data and compute the average value. Print the number of data values and the average value.

SOLUTION

This example resembles Example 4-1, except that we need to initialize the counter to zero and end the loop when we reach the trailer signal.

Pseudocode

```
solution_2:    sum ← 0
               count ← 0
               read time, temperature
               while time ≠ − 999
                       add temperature to sum
                       add 1 to count
                       read time, temperature
               if count = 0
                       print error message
               else
                       average ← sum/count
                       print count, average
```

Fortran Program

```fortran
!-------------------------------------------------------------------!
program solution_2
!
!   This program computes the average temperature
!   using a set of times and temperatures stored in
!   a data file. The data file contains a trailer
!   line containing -999 for time and temperature.
!
!
!   Declare and initialize variables.
    implicit none
    integer :: count=0
    real :: sum=0.0, temperature, time
!
!   Open file and read first set of values.
    open(unit=10,file="results2.dat",status="old")
    read(10,*) time, temperature
!
!   Compute temperature sum until trailer line.
    do while (time /= -999)
       sum = sum + temperature
       count = count + 1
       read(10,*) time, temperature
    end do
!
!   Print average.
    if (count == 0) then
       print*, "No data values in this file"
    else
       print 25, count, sum/real(count)
 25    format (1x,"count = ",i3,5x,"average = ",&
               f8.2," degrees F")
    end if
```

```
!
!     Close file.
      close(unit=10)
!
end program solution_2
!-------------------------------------------------------------------!
```

. . . .

end **Option**

If we do not know the number of data lines in a file, and if no trailer signal appears at the end, we must use a different technique. The following are the steps, in pseudocode, that we want to perform:

> while there is more data
> read variables
> process variables

To implement these instructions in Fortran, we use an option available with the read statement. This option tests for the end of the data and branches to a specified statement if the end is detected. A list-directed read statement that uses this option has the following general form:

> read(*unit number,*,*end=*n) *variable list*

As long as there is data to read in the data file, this statement executes exactly like this statement:

> read(*unit number,*) variable list*

However, if the last line of the data file has already been read and we execute the read statement with the *end option,* then instead of getting an execution error, control is passed to the statement referenced in the end option. If the read statement is executed a second time after the end of the data has been reached, an execution error occurs.

A read statement with the end option is actually a special form of the while loop. When using it in a program, it should be used inside a do loop generated by a do statement with no parameters listed on it, as shown below:

```
        integer :: count=0
        real :: sum=0, temperature, time
        ...
        do
           read(10,*,end=15) time, temperature
           sum = sum + temperature
           count = count + 1
        end do
15      ...
```

This special implementation of the while loop should be used only when you do not know the number of data lines to be read and no trailer signal appears at the end of the file. If you have a trailer signal, test for that value to exit the loop instead of using the end option. The choice of the correct technique for reading data from a data file depends on the information in the data file. We can now solve the problem of computing an average of laboratory data for a file with no trailer signal and no way to determine the number of records in the file before running the program.

EXAMPLE 4-3

Average of Laboratory Data

Assume that we have a data file named `results3.dat` that contains information collected from a laboratory experiment. Each line of the data file contains two numbers: a time value in seconds and a temperature measurement in degrees Fahrenheit. There is no special information in the first record containing the number of lines in the file, and there is no trailer signal. Read the data and compute the average value. Print the number of data values and the average value.

SOLUTION

In this example we use the end option to branch to the statement that performs the calculations.

Pseudocode

```
solution_3:    sum ← 0
               count ← 0
               while more data
                       read time, temperature
                       add temperature to sum
                       add 1 to count
               if count = 0 then
                       print error message
               else
                       average ← sum/count
                       print count, average
```

Fortran Program

```fortran
!-----------------------------------------------------------!
program solution_3
!
!  This program computes the average temperature
!  using a set of times and temperatures stored in
!  a data file. The data file contains only the
!  time and temperature values.
!
!
!     Declare and initialize variables.
      implicit none
      integer :: count=0
      real :: sum=0.0, temperature, time
!
!     Open file.
      open(unit=10,file="results3.dat",status="old")
!
!     Compute temperature sum until
!     the end of the data is reached.
```

```
        do
            read(10,*,end=15) time, temperature
            sum = sum + temperature
            count = count + 1
        end do
    !
    !    Print average.
    15 if (count == 0) then
            print*, "No data values in this file"
        else
            print 25, count, sum/real(count)
    25      format (1x,"count = ",i3,5x,"average = ",&
                    f8.2," degrees F")
        end if
    !
    !    Close file.
        close(unit=10)
    !
    end program solution_3
    !-------------------------------------------------------------------!
```

. . . .

Try It

Try this self-test to check your memory of some key points from Section 4-2. If you have any problems with the exercises, you should reread this section. The solutions are given at the end of this module.

Assume that a data file contains time and altitude measurements from a rocket trajectory. Each line of the data file contains a corresponding time and altitude measurement. A trailer signal with a time equal to −99 is used to signal the end of the data file. Assume that the program contains the following statements:

```
real :: altitude, time
...
read(10,*) time, altitude
do while (time /= -99)
    print*, time, altitude
    read(10,*) time, altitude
end do
```

For each of the following data lines, indicate if the line would work properly as the trailer line. If the line would not work properly, explain why it would not work.

1. −99 −99
2. −99 100
3. 100 −99
4. −99
5. −99 100 15.7

4-3 Application   CRITICAL PATH ANALYSIS

Manufacturing Engineering

A critical path analysis is a technique used to determine the time schedule for a project. This information is important in the planning stages before a project is begun, and it is also useful to evaluate the progress of a project that is partially completed. One method for this analysis starts by breaking a project into sequential events, then breaking each event into various tasks. Although one event must be completed before the next one is started, various tasks within an event can occur simultaneously. The time it takes to complete an event, therefore, depends on the number of days required to finish its longest task. Similarly, the total time it takes to finish a project is the sum of the times it takes to finish the individual events.

Assume that the critical path information for a major construction project has been stored in a data file in the computer. You have been asked to analyze this information so that your company can develop a bid on the project. Specifically, the management would like a summary report that lists each event along with the number of days for the shortest task in the event and the number of days for the longest task in that event. In addition, they would like the total project length computed in days and converted to weeks (five days per week).

The data file named `project.dat` contains three integers per line. The first number is the event number, the second number is the task number, and the third number is the number of days required to complete the task. The data have been stored such that all the task data for event 1 is followed by all the task data for event 2, and so on. The task data for a particular event is also in order. You do not know ahead of time how many entries are in the file, but there is an upper limit of 98 total events, so a trailer signal is used to indicate the last line in the data file. The last line will contain a value of 99 for the event number and zeros for the corresponding task and days.

1. Problem Statement

Determine and print a project completion timetable.

2. Input/Output Description

Input—a data file named `project.dat` that contains the critical path information for the events in the project

Output—a report

3. Hand Example

Use the following set of project data for the hand example:

Event Number	Task Number	Days
1	1	5
1	2	4
2	1	3
2	2	7
2	3	4
3	1	6
99	0	0

The corresponding report based on this data is as follows:

```
Project Completion Timetable

Event Number  Task Minimum  Task Maximum
     1             4             5
     2             3             7
     3             6             6

Total Project Length = 18 days
                     =  3 weeks 3 days
```

4. Algorithm Development

Decomposition

Read project event data and print event information.
Print summary information.

To develop an algorithm, we look at the way we compiled the data in the hand example. Because all the data for the first event is together, we scan down the list of data for event 1 and keep track of the minimum number of days and maximum number of days. When we reach the task for event 2, we print the information related to event 1 and then repeat the process for event 2. We continue until we reach event 99, which is a signal that we do not have any more data. We print the information for the last event and then print the summary information using a total in which we accumulated the total maximum number of days for all the events.

Initial Pseudocode

```
critical_path:    total ← 0
                  read event, task, days
                  number ← event
```

```
            min_days ← days
            max_days ← days
            while not done
                  if same event then
                        update min_days and max_days
                  else
                        print number, min_days, max_days
                        add max_days to total
                        number ← event
                        min_days ← days
                        max_days ← days
                  read event, task, days
            print number, min_days, max_days for last event
            add max_days to total
            print total
```

As we refine the pseudocode, we need to be more specific about how we determine if we are done. We also need to add more details to the update for the minimum and maximum days for an event. Finally, when we print the total, we also need to compute and print the number of weeks.

Final Pseudocode

```
critical_path:    total ← 0
                  read event, task, days
                  number ← event
                  min_days ← days
                  max_days ← days
                  if event = 99 then
                        done ← true
                  else
                        done ← false
                  while not done
                        if event = number then
                              if days < min_days then min_days ← days
                              if days > max_days then max_days ← days
                        else
                              print number, min_days, max_days
                              add max_days to total
                              number ← event
                              min_days ← days
                              max_days ← days
                        read event, task, days
                        if event = 99 then done ← true
                  print number, min_days, max_days for last event
                  add max_days to total
                  weeks ← total/5
                  plus ← total − weeks·5
                  print total, weeks, plus
```

Note that we need to do the computation for weeks so that the result is an integer instead of a real value.

Fortran Program

```
!------------------------------------------------------------------!
program critical_path
!
!   This program determines the critical path information
!   from project data stored in a data file.
!
!
!     Declare and initialize variables.
      implicit none
      integer :: days, event, max_days, min_days, &
                 number, plus, task, total=0, weeks
      logical :: done
!
!     Open data file.
      open(unit=8,file="project.dat",status="old")
!
!     Print headers.
      print*, "Project Completion Timetable"
      print*
      print*, "Event Number      Task Minimum      Task Maximum"
!
!     Read initial set of data values.
      read(8,*) event, task, days
      number = event
      min_days = days
      max_days = days
      if (event == 99) then
         done = .true.
      else
         done = .false.
      end if
!
!     Read the rest of data and generate report.
      do while (.not.done)
         if (event == number) then
            if (days < min_days) then
               min_days = days
            else if (days > max_days) then
               max_days = days
            end if
         else
            print 10, number, min_days, max_days
   10       format(1x,i6,11x,i6,11x,i6)
            total = total + max_days
            number = event
            min_days = days
            max_days = days
         end if
         read(8,*) event, task, days
         if (event == 99) done = .true.
      end do
```

```
!
!     Print final event data.
      print 10, number, min_days, max_days
      total = total + max_days
!
!     Print final project length summary.
      weeks = total/5
      plus = total - weeks*5
      print 15, total
   15 format(/,1x,"Total Project Length = ",i2," days")
      print 20, weeks, plus
   20 format(1x,20x," = ",i2," weeks",i2," days")
!
!     Close file.
      close(file=8)
!
end program critical_path
!-------------------------------------------------------------!
```

The data files used in the Application sections, along with the corresponding programs, are available on a diskette that is available to your instructor. Check with your instructor to see if this information has been stored on your computer system so that you can access these programs and data files.

5. Testing

If we use the sample set of data from the hand example with this program, the following report is printed:

```
Project Completion Timetable

Event Number        Task Minimum        Task Maximum
     1                   4                   5
     2                   3                   7
     3                   6                   6

Total Project Length = 18 days
                     =  3 weeks 3 days
```

4-4 Application   **SONAR SIGNALS**

Acoustical Engineering

The study of sonar (sound navigation and ranging) includes the generation, transmission, and reception of sound energy in water. Oceanological applications include ocean topography, geological mapping, and biological signal measurements; industrial sonar applications include fish finding, oil and mineral exploration, and acoustic beacon navigation; naval sonar applications include submarine navigation and submarine tracking. An active sonar system transmits a signal that often is a cosine with a known frequency. The reflections or echoes of the signal are then received and analyzed to provide information about the surroundings. A passive sonar system does not transmit signals but collects signals from sensors and analyzes them based on their frequency content.

To test algorithms that analyze sonar signals, engineers and scientists would first work with simple simulated sonar signals instead of actual sonar signals. A simple sonar signal can be represented by the following equation:[1]

$$x(t) = \sqrt{\frac{2E}{PD}} \cos(2\pi f_c t) \qquad ,0 \le t \le PD$$

$$= 0 \qquad\qquad ,\text{otherwise}$$

where   E = the transmitted energy in joules
PD = the pulse duration in seconds
f_c = the frequency in Hz
t = the time variable in seconds

Durations of a sonar signal can range from a fraction of a millisecond to several seconds, and frequencies can range from a few hundred Hz (Hertz) to tens of kHz (kiloHertz), depending on the sonar system and its desired operating range.

Write a program in which the user inputs the values for E, PD, and f_c. Then generate a data file called sonar1.dat that contains samples of this sonar signal. The sampling of the signal should cover the pulse duration and be such that there are ten samples for every period of x(t), where a period is equal to $1/f_c$ second. The data file should contain two values for every line: a time value (starting at zero) and the corresponding value of the sonar signal.

1. Problem Statement

Generate a sonar signal that contains ten samples from each period of a specified cosine, covering a given time duration.

2. Input/Output Description

Input—the values of E (transmitted energy in joules), PD (pulse duration in seconds), and f_c (frequency in Hz)

Output—a data file named sonar1.dat that contains time and signal values for the sonar pulse duration

---

[1] A. B. Baggeroer, ''Sonar Signal Processing,'' Chapter 6 of *Applications of Digital Signal Processing* (Englewood Cliffs, N.J.: Prentice-Hall, 1978).

3. Hand Example

For a hand example, we use the following values:

$$E = 500 \text{ joules}$$
$$PD = 0.5 \text{ millisecond (ms)}$$
$$f_c = 3.5 \text{ kHz}$$

The period of the cosine is 1/3500, or approximately 0.3 ms. Thus, to have 10 samples per period, the sampling interval must be approximately 0.03 ms. The pulse duration is 0.5 ms, and thus we need 17 samples of the following signal:

$$x(t) = \sqrt{\frac{2E}{PD}} \cos(2\pi f_c t)$$
$$= \sqrt{\frac{1000}{0.0005}} \cos(2\pi 3500t)$$
$$= 1414.2 \cos(2\pi 500t)$$

The values for the sonar signal are

t (ms)	x(t)
0.00	1414.2
0.03	1117.4
0.06	351.7
0.09	−561.6
0.12	−1239.3
0.15	−1396.8
0.18	−968.1
0.21	−133.1
0.24	757.7
0.27	1330.6
0.30	1345.0
0.33	794.9
0.36	−88.8
0.39	−935.2
0.42	−1389.2
0.45	−1260.1
0.48	−602.1

4. Algorithm Development

Decomposition

Read energy, pulse duration, frequency.
Compute cosine amplitude, sampling time, number of points.
Generate file containing time values and cosine values.

The problem description did not specify using a trailer signal or an initial record with the number of lines in the data file; therefore, we will generate a data file with only valid data lines.

Pseudocode

sonar_signal: read energy (E), pulse duration (PD), frequency

$$A \leftarrow \sqrt{\frac{2E}{PD}}$$

$$period \leftarrow \frac{1}{frequency}$$

$$sample\ interval \leftarrow \frac{period}{10}$$

$$number\ of\ samples \leftarrow \frac{PD}{sample\ interval}$$

$k \leftarrow 0$
while $k \leq$ (number of samples $- 1$)
    time $\leftarrow k \cdot$ (sample interval)
    signal $\leftarrow A \cos(2\pi \cdot$ frequency $\cdot$ time)
    write time and signal to data file
    add 1 to k

Fortran Program

```
!----------------------------------------------------------------!
program sonar_signal
!
!   This program generates a sonar signal.
!
!
!     Declare and initialize variables.
      implicit none
      integer :: k, number
      real :: A, E, frequency, PD, pi, period, T, &
              time, signal
      parameter (pi=3.141593)
!
!     Open output file.
      open(unit=10,file="sonar1.dat",status="unknown")
!
!     Prompt user for input information.
      print*, "Enter transmitted energy in joules:"
      read*, E
      print*, "Enter pulse duration in seconds:"
      read*, PD
      print*, "Enter cosine frequency in Hertz:"
      read*, frequency
!
!     Compute signal parameters.
      A = sqrt(2.0*E/PD)        ! signal amplitude
      period = 1.0/frequency    ! signal period
      T = period/10.0           ! sample time interval
      number = nint(PD/T)       ! number of samples to compute
```

```
!
!       Generate signal values and write to a data file.
        do k=0,number-1
            time = real(k)*T
            signal = A*cos(2.0*pi*frequency*time)
            write(10,*) time, signal
        end do
!
!       Close file.
        close(unit=10)
!
end program sonar_signal
!-----------------------------------------------------------!
```

The nearest integer function (nint) was used in the computation of the number of samples to round the answer. If a file sonar1.dat exists when this program is executed, the new file will replace the existing file since the file status is specified as "unknown"; otherwise, if the file status were specified to be "new", we would need to delete any existing file named sonar1.dat before executing the program to avoid an execution error.

5. Testing

The following are the terminal screen interaction and a plot of the corresponding sonar1.dat file generated by MATLAB:

```
Enter transmitted energy in joules:
500.0
Enter pulse duration in seconds:
0.005
Enter cosine frequency in Hertz:
3500.0
```

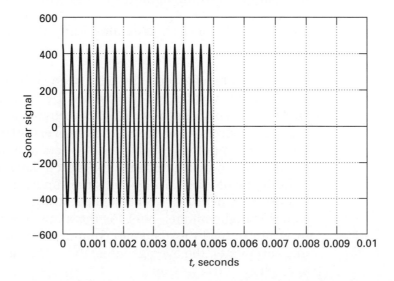

SUMMARY

Data files are necessary in engineering and scientific applications involving large amounts of data. This chapter presented the I/O statements for working with data files. It also discussed how to read information from data files. In particular, the chapter presented three techniques for reading data files that depend on whether the data file includes a trailer signal or an initial line that contains the number of valid data records that follow it. We will use data files frequently throughout this module.

Key Words

data file	sentinel signal
end option	trailer signal
record	

Problems

This problem set begins with modifications to programs developed earlier in this chapter. Give the decomposition, pseudocode or flowchart, and Fortran program for each problem.

Problem 1 modifies the program `critical_path`, given in Section 4-3 on page 110.

1. Modify the critical path analysis program so that it prints the event and task number for all tasks requiring more than five days.

Problems 2–3 modify the program `sonar_signal`, given in Section 4-4 on page 114.

2. Modify the sonar signal program so that the output data file has a trailer signal with − 99 as both numbers on a trailer line.

3. Modify the sonar signal program so that it prints a first line in the data file that contains the number of valid data lines that follow.

In problems 4–12 develop programs using the five-step problem-solving process.

4. As a practicing engineer, you have been collecting data on the performance of a new solar device. You have been measuring the sun's intensity and the voltage produced by a photovoltaic cell exposed to the sun. These measurements have been taken every 30 minutes during daylight hours for 2 months. Because the sun sets at a different time each day, the number of measurements taken each day may vary.

    Each line of valid data contains the sun's intensity (integer), the time (24 hours, where 1430 represents 2:30 pm), and the voltage (real). A trailer signal contains 9999 for the sun's intensity. Write a complete program to read this data and compute and print the total number of measurements, the average intensity of the sun, and the average voltage value.

5. Write a program to compute fuel cost information for an automobile. The input data is stored in a data file called `miles.dat`. The first line in the file contains the initial real mileage (odometer reading). Each following data line contains information collected as the automobile was refueled and includes the new mileage reading, the cost of the fuel, and the number of gallons of fuel. (All the values are real values.) The last line of the file contains a negative value and two zeros, instead of the refueling information. The output of the program should be the cost per mile and the number of miles per gallon for each line of input data. Use the following report format:

```
Fuel Cost Information
Miles   Gallons   Cost   Cost/Mile   Miles/Gallon
xxx.x   xx.x      xx.xx  xx.xx       xx.x
```

6. Add the following summary information at the end of the report printed in problem 5.

```
Summary Information
Total Miles           xxxx.x
Total Cost            xxxx.xx
Total Gallons         xxxx.xx
Average Cost/Mile     x.xx
Average Miles/Gallon  xx.x
```

7. Oil exploration and recovery is an important concern of large petroleum companies. Profitable oil recovery requires careful testing by drilling seismic holes and blasting with specific amounts of dynamite. Optimum seismic readings require a specific ratio between the amount of dynamite and the depth of the hole. Assume that each stick of dynamite is 2.5 feet long and weighs 5 pounds. The ideal powder charge requires a dynamite to depth-of-hole ratio of 1:3. Thus, a 60-foot hole would require 20 feet of dynamite, which is equal to 8 sticks, or 40 pounds. The actual powder charge is not always equal to ideal powder charge because the ideal powder charge may not be in 5-pound increments; in these cases, the actual powder charge should be rounded down to the nearest 5-pound increment. (You cannot cut or break the dynamite into shorter lengths for field operations.) The following example should clarify this process:

Hole depth = 85 feet
Ideal charge = 85/3   = 28.3333 feet
                     = 11.3333 sticks
                     = 56.6666 pounds
Actual charge = 55 pounds
              = 11 sticks

Information on the depths of the holes to be tested each day is stored in a data file called `drill.dat`. The first line contains the number of sites to be tested that day. Each following line contains integer information for a specified site that gives the site identification number and the depth of the hole in feet. Write a complete program to read this information and print a report in the following format:

```
Daily Drilling Report
Site     Depth     Ideal Powder     Actual Powder     Sticks
ID       (ft)      Charge (lbs)     Charge (lbs)
12980    85        56.6666          55                11
```

8. Modify the program in problem 7 so that a final summary report follows the drilling report and has the form

```
Total Powder Used = xxxxx lbs    (xxxx sticks)
Total Drilling Footage = xxxxxx ft
```

9. Modify the program in problem 8 so that it takes into consideration a special situation: If the depth of the hole is less than 30 feet, the hole is too shallow for blasting. Instead of printing the charge values for such a hole, print the site identification number, the depth, and the message Hole too shallow for blasting. The summary report should not include data for these shallow holes. Add a line to the summary report that contains the number of holes too shallow for blasting.

10. Assume that a data file called dataxy.dat contains a set of data coordinates (real values) that are to be used by several different programs. The first line of the file contains the number of data coordinates in the file, and each of the remaining lines contains the x and y coordinates for one of the data points. Since some of the programs need polar coordinates, generate a second data file that has each point in polar form, which is a radius and an angle in radians. Then, no matter how many programs use the data, it only has to be converted to polar coordinates once, and each program can then reference the appropriate file. Write a program to generate a new file called polar.dat that contains the coordinates in polar form instead of rectangular form. The following equations convert a coordinate in rectangular form to polar form:

$$r = \sqrt{x^2 + y^2} \qquad \theta = \arctan\left(\frac{y}{x}\right)$$

(Be sure that the first line of the new data file specifies the number of data coordinates.)

11. Rewrite the program from problem 10, assuming that the original file is polar.dat and that it contains data coordinates in polar form. The new output file should be called dataxy.dat and should contain data coordinates (real values) in rectangular form. The equations for converting polar coordinates to rectangular coordinates are

$$x = r\cos(\theta) \qquad y = r\sin(\theta)$$

5 Array Processing

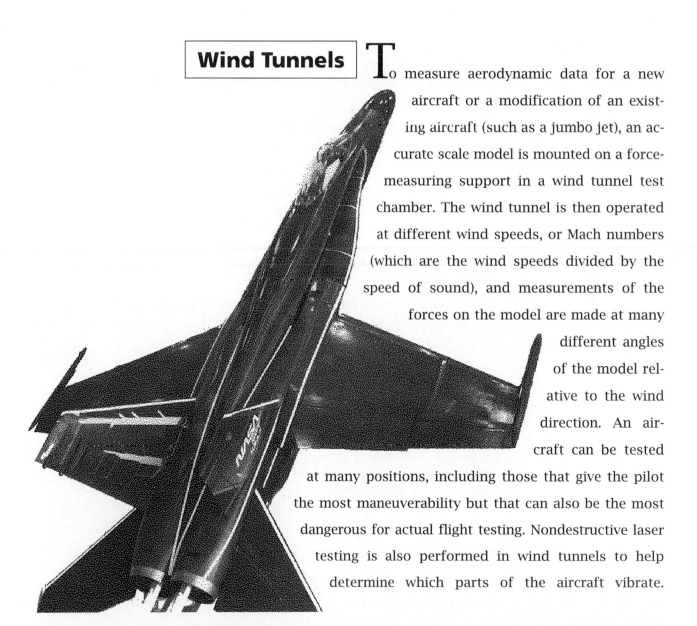

Wind Tunnels To measure aerodynamic data for a new aircraft or a modification of an existing aircraft (such as a jumbo jet), an accurate scale model is mounted on a force-measuring support in a wind tunnel test chamber. The wind tunnel is then operated at different wind speeds, or Mach numbers (which are the wind speeds divided by the speed of sound), and measurements of the forces on the model are made at many different angles of the model relative to the wind direction. An aircraft can be tested at many positions, including those that give the pilot the most maneuverability but that can also be the most dangerous for actual flight testing. Nondestructive laser testing is also performed in wind tunnels to help determine which parts of the aircraft vibrate.

INTRODUCTION

This chapter develops a method for storing groups of values without explicitly giving each value a different name. Each group (called an array) has a common name, but each individual value has a unique index or subscript. This technique allows us to analyze the data using loops, where the common name remains the same but the index or subscript becomes a variable that changes with each pass through the loop. Because the data values are stored in separate memory locations, we can also access the data as often as needed without rereading it or recomputing it.

5-1 ONE-DIMENSIONAL ARRAYS

An *array* is a group of storage locations that have the same name and data type. Each storage location in an array is called an *element* and is distinguished by using the common name followed by a *subscript* or an index in parentheses. Subscripts are represented by consecutive integers, usually beginning with the integer 1. A *one-dimensional array* can be visualized as either one column of data or one row of data. The following diagram shows the storage locations and associated names for a one-dimensional integer array j of five elements and a one-dimensional real array distance with four elements:

j(1)	j(2)	j(3)	j(4)	j(5)
2	-5	14	80	-12

distance(1)	1.2
distance(2)	-0.8
distance(3)	36.9
distance(4)	-0.07

Array Storage

Memory space is reserved for an array by specifying the array name and size on a type statement. For example, the arrays j and distance above are defined with the following statements:

```
integer :: j(5)
real :: distance(4)
```

Array sizes must be specified with constants, not variables; however, an array size can be specified with a variable that has been defined using the parameter statement that was discussed in Chapter 2. An alternative definition uses a dimension statement to specify the size of the array separately from the type specification:

```
integer :: j
real :: distance
dimension :: distance(4), j(5)
```

The typing of an array, whether implicit or explicit, applies to all elements of the array; hence, an array cannot contain both real values and integer values. Because explicit typing is desirable, we will use explicit typing statements instead of `dimension` statements for defining the arrays in the examples in this module.

The range of subscripts associated with an array can be specified with a beginning subscript number and an ending subscript number. Both numbers must be integers separated by a colon and must follow the array name in the type statement or the `dimension` statement. The following statements reserve storage for a real array `tax` whose elements are `tax(0)`, `tax(1)`, `tax(2)`, `tax(3)`, `tax(4)`, and `tax(5)` and an integer array `income` whose elements are `income(-3)`, `income(-2)`, `income(-1)`, `income(0)`, `income(1)`, `income(2)`, and `income(3)`:

```
integer :: income(-3:3)
real :: tax(0:5)
```

Also note that the following two declarations are equivalent:

```
integer :: area(1:15)
integer :: area(15)
```

Unless otherwise stated, we will assume in this module that all array subscripts begin with the integer 1. However, sometimes the range of the subscripts logically starts with an integer other than 1. For example, if we have a set of population values for the years 1880 through 1980, it might be convenient to use the year to specify the corresponding population. We could specify such an array with the following statement:

```
integer :: population(1880:1980)
```

Then, if we wish to refer to the population for 1885, we use the reference `population(1885)`.

Initialization and Computations

Values are assigned to array elements in the same way that values are assigned to regular variables. The following are valid assignment statements if `number`, `count`, and `x` have been assigned values:

```
integer :: count, j(5), number
real :: distance(4), x
...
j(1) = 0
j(5) = number*count
distance(2) = 46.2/sin(x)
```

An array name without a subscript can be used in an assignment statement. For example, the following statements initialize all the elements of the array `j` to zero and all of the elements of the array `distance` to 1.5:

```
integer :: j(5)
real :: distance(4)
...
j = 0
distance = 1.5
```

If all the elements of an array are to be assigned the same value, this assignment can also occur on the type statement, as shown with the following equivalent statements:

```
integer :: j(5)=0
real :: distance(4)=1.5
```

Variables and expressions can also be used as subscripts. For example, the following loop initializes the elements of the array j to the integer values 1 through 5. Observe that the variable i is used as a subscript and also as the do loop index:

```
integer :: i, j(5)
...
do i=1,5
   j(i) = i
end do
```

j(1)	j(2)	j(3)	j(4)	j(5)
1	2	3	4	5

The values of the array distance are initialized to real values with this set of statements:

```
integer :: k
real :: distance(4)
...
do k=1,4
   distance(k) = real(k)*1.5
end do
```

distance(1)	distance(2)	distance(3)	distance(4)
1.5	3.0	4.5	6.0

The previous examples illustrate that a subscript can be an integer constant or an integer variable. Subscripts can also be integer expressions, as indicated in the following statements, which assume that j, r, and trace have been defined as integer arrays, and that i, k, and b have been defined as integers:

```
j(2*i) = 3
r(k) = r(k-1)
b = trace(2*i) + trace(2*i+1)
```

Whenever an expression is used as a subscript, be sure the value of the expression is between the starting and ending subscript values. If a subscript is outside the proper range, the program will not work correctly. With some compilers, a logic error message is given if a subscript is out of bounds; other compilers use an incorrect value for the invalid array reference, causing serious errors that are difficult to detect.

If an array name is used without a subscript in an expression or in an assignment statement, the statements are interpreted element-by-element.

To illustrate, assume that a, b, and c are real arrays with ten elements. Then, the following assignment statement causes 2.5 to be added to each element in the array b and is equivalent to the loop on the right:

```
real :: b(10)           integer :: k
...                     real :: b(10)
b = b + 2.5             ...
                        do k=1,10
                            b(k) = b(k) + 2.5
                        end do
```

The next assignment statement specifies that corresponding elements of a and b are to be added and stored in the corresponding elements of c. An equivalent loop is also shown on the right:

```
real :: a(10), b(10), c(10)      integer :: k
...                              real :: a(10), b(10), c(10)
c = a + b                        ...
                                 do k=1,10
                                     c(k) = a(k) + b(k)
                                 end do
```

In the following assignment statement and equivalent loop, each element in the array a is replaced by the tangent of its value:

```
real :: a(10)           integer :: k
...                     real :: a(10)
a = tan(a)              ...
                        do k=1,10
                            a(k) = tan(a(k))
                        end do
```

Input and Output

To read data into an array from a keyboard or from a data file, we use the read statement. If we wish to read an entire array, we can use the name of the array without subscripts. We can also specify specific elements in a read statement. If the array a contains three elements, then the following two read statements are equivalent; if the array a contains eight elements, then the first read statement reads values for all eight elements and the second read statement reads values for only the first three elements:

```
read*, a
read*, a(1), a(2), a(3)
```

Array values may also be read using an *implied do loop.* Implied do loops use the indexing feature of the do statement and may be used only in input and output statements and in the data statement, which is presented later in this section. For example, if we want to read the first ten elements of the array r, we can use the following implied do loop in the read statement:

```
read*, (r(i),i=1,10)
```

Further examples illustrate the use of these techniques for reading data into an array.

EXAMPLE 5-1   **Temperature Measurements**

A set of 50 temperature measurements has been entered into a data file, one value per line. The file is accessed with unit number 9. Give a set of statements to read this data into an array:

```
temperature(1)    Line 1 of data file
temperature(2)    Line 2 of data file
...
temperature(50)   Line 50 of data file
```

SOLUTION 1

```
integer :: i
real :: temperature(50)
...
do i=1,50
    read(9,*) temperature(i)
end do
```

The read statement in this solution reads one value, but it is in a loop that is executed 50 times, and thus it reads the entire array.

SOLUTION 2

```
real :: temperature(50)
...
read(9,*) temperature
```

The read statement in this solution contains no subscript; thus, it reads the entire array.

SOLUTION 3

```
integer :: i
real :: temperature(50)
...
read(9,*) (temperature(i),i=1,50)
```

The read statement in this solution contains an implied loop and is equivalent to a read statement that lists temperature(1), temperature(2), . . . , temperature(50).

Note that solutions 2 and 3 are exactly the same as far as the computer is concerned; they both represent one read statement with 50 variables. Solution 1 is the same as 50 read statements with 1 variable per read statement. If the data file contains 50 lines, each with 1 temperature measurement, all three solutions store the same data in the array temperature.

Suppose that each of the 50 lines in the data file has two numbers: a temperature measurement and a humidity measurement. Solution 1 will read a new data line for each temperature measurement because it is the equivalent of 50 read statements. But because solutions 2 and 3 are the equivalent of one read statement with 50 variables listed, they will go to a new line only when they run out of data. Thus, the data is stored as shown:

```
temperature(1)   First temperature
temperature(2)   First humidity
temperature(3)   Second temperature
temperature(4)   Second humidity
...
```

This is a subtle but important distinction: The computer does not recognize that an error has occurred in the last example shown. It has data for the array and continues processing, assuming it has the correct data.

. . . .

EXAMPLE 5-2   ## Mass Measurements

A group of 30 mass measurements is stored in a real array `mass`. Print the values in the following tabulation:

```
mass( 1) = xxx.x kg
mass( 2) = xxx.x kg
...
mass(30) = xxx.x kg
```

SOLUTION

```
integer :: i
real :: mass(30)
...
do i=1,30
    print 15, i, mass(i)
15  format(1x,"mass(",i2,") = ",f5.1," kg")
end do
```

For each output line, we need to reference one value in the array. We can generate the values of the subscript with a do statement. The output form of this solution is important; go through it carefully to be sure you understand the placement of the literals in the format statement.

. . . .

Try It   Try this self-test to check your memory of some key points in this material. If you have any problems with the exercises, you should reread this section. The solutions are given at the end of this module.

Problems 1–3 contain statements that initialize and print one-dimensional arrays. Show the output from each set of statements. Assume that each set of statements is independent of the others.

1.
```
    integer :: i, m(10)
    ...
    do i=1,10
        m(i) = 11 - i
    end do
    print*, "array values:"
    do i=1,10
        print 10, i, m(i)
10      format(1x,"m(",i2,") = ",i4)
    end do
```

2.
```
    integer :: k, list(8)
    ...
    do k=1,8
        list(k) = k**2 - 4
    end do
    print 10, list
10  format(1x,8i4)
```

```
3.      integer :: j
        real :: time(20)
        ...
        time(1) = 3.0
        do j=2,20
           time(j) = time(j-1) + 0.5
        end do
        print 10, (j, time(j), j=1,20,4)
  10 format(1x,"time ",i2," = ",f5.2)
```

Intrinsic Functions

Fortran 90 contains intrinsic functions specifically written to work with arrays. Table 5-1 contains a brief description of some of these functions. Additional array functions are included in the expanded listing of intrinsic functions in Appendix A.

Table 5-1   Intrinsic Functions for Working with Arrays

Function Name and Argument	Function Value
minval(x)	Determines the minimum value in the array x
maxval(x)	Determines the maximum value in the array x
product(x)	Computes the product of the values in the array x
sum(x)	Computes the sum of the values in the array x

Try It

Try this self-test to see if you understand the array functions. If you have any problems with the exercises, you should reread the material in this section. The solutions are given at the end of this module.

For each of the problems below, assume that the integer array a has been defined and has the following contents:

a(1)	a(2)	a(3)	a(4)	a(5)	a(6)
−2	8	5	−4	3	7

Give the output of each statement:

1. print*, "Sum = ", sum(a)
2. print*, product(a), "is the product of elements"
3. print*, "Maximum: ", maxval(a)
   print*, "Minimum: ", minval(a)

Data Statement

The data statement is a specification statement and is therefore nonexecutable; it is useful in initializing both simple variables and arrays. The general form of the data statement is

> data *list of variable names / list of constants /*

An example of a `data` statement to initialize simple variables is

```
real :: length, sum, velocity, voltage
data length, sum, velocity, voltage &
     /10.0, 0.0, 32.75, -2.5/
```

The number of data values must match the number of variable names; the data values should also be of the correct type. The preceding `data` statement initializes the following variables:

| length | 10.0 | sum | 0.0 |
| velocity | 32.75 | voltage | − 2.5 |

Because the `data` statement is a specification statement, it should precede any executable statements; it is therefore located near the beginning of your program, along with the `real`, `integer`, `logical`, and `dimension` statements. It follows these other statements because the specification of the types of variables or the declaration of an array should precede values given to the corresponding memory locations.

Caution should be exercised when using the `data` statement because it initializes values only at the beginning of program execution; this means that the `data` statement cannot be used in a loop to reinitialize variables. If it is necessary to reinitialize variables, you must use assignment statements.

When a number of values are to be repeated in a list of values, a constant followed by an asterisk is used to indicate the repetition. The following statement initializes all four variables to zero:

```
real :: a, b, c, d
data a, b, c, d /4*0.0/
```

A `data` statement can also be used to initialize one or more elements of an array:

```
integer :: j(5)
real :: time(4)
data j, time /5*0, 1.0, 2.0, 3.0, 4.0/
```

| j | 0 | 0 | 0 | 0 | 0 |

| time | 1.0 | 2.0 | 3.0 | 4.0 |

```
real :: hours(5)
data hours(1) /60.0/
```

| hours | 60.0 | ? | ? | ? | ? |

The question marks indicate that some array elements were not initialized by the `data` statement. A syntax error would have occurred if the subscript was left off the array reference hours(1), because the number of variables would then not match the number of data values; that is, hours represents five variables, but hours(1) represents only one variable.

An implied loop can also be used in a data statement to initialize all or part of an array, as in

```
integer :: year(100)
data (year(i), i=1,50) /50*0/
```

The first 50 elements of the array are initialized to zero, and the last 50 are not initialized. You must therefore make no assumptions about the contents of the last 50 values in the array.

5-2 Application   WIND TUNNELS

Aerospace Engineering

Wind tunnels are used to collect very accurate measurements of the lift and drag forces generated by a moving air stream on an aerodynamic body. When the data are subsequently used in engineering analyses, linear interpolation is frequently used to estimate values between the measured data points. This section explains linear interpolation and then uses linear interpolation in the analysis of data from a wind tunnel.

Linear Interpolation

Interpolation is a numerical technique used to estimate unknown values of a function by using known values of the function at specific points. The simplest and probably the most popular interpolation method is linear interpolation. Essentially, all that is involved is connecting the data points with straight line segments and then interpolating for unknown values of the function on the straight line segments that connect adjacent data points. For example, consider the following graph that contains two values of a function at x = 1 and at x = 3. These two points have been connected with a straight line.

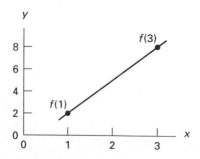

We can use linear interpolation to estimate the value of the function at x = 1.5 using similar triangles, as shown in the following figure:

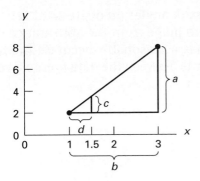

Thus,

$$\frac{a}{b} = \frac{c}{d}$$

or

$$\frac{f(3) - f(1)}{3 - 1} = \frac{f(1.5) - f(1)}{1.5 - 1}$$

When we solve this equation for f(1.5), we have

$$f(1.5) - f(1) = \frac{(1.5 - 1) \cdot (f(3) - f(1))}{3 - 1}$$

or

$$f(1.5) = f(1) + \frac{(1.5 - 1) \cdot (f(3) - f(1))}{3 - 1}$$

$$= 2 + \frac{(0.5)(8 - 2)}{2}$$

$$= 3.5$$

We can develop a general linear interpolation formula for finding a function value f(x) when x is between a and b and when the values of f(a) and f(b) are known. This formula is essentially the formula used in the previous example. It calculates the portion of the difference between the values on the x axis and then adds the corresponding portion of the difference between the values on the y axis to the leftmost function value:

$$f(x) = f(a) + \frac{(x - a)}{(b - a)} (f(b) - f(a))$$

This general formula works for linear interpolation of straight lines with either positive or negative slopes.

Computation of Lift and Drag Forces

In this section we use a set of lift data on an aircraft wing. The 17 points in the data file are tabulated and plotted to show the coefficient of lift,

C_L, at various flight path angles (in degrees) of the wing for a Mach number of 0.3. It is easy to judge from the appearance of the data that linear interpolation provides a reasonably accurate method of estimating the lift coefficient at points between the data points, even in the region where the function is curved.

Flight Path Angle	Coefficient of Lift
−4	−0.202
−2	−0.500
0	0.108
2	0.264
4	0.421
6	0.573
8	0.727
10	0.880
12	1.027
14	1.150
15	1.195
16	1.225
17	1.244
18	1.250
19	1.245
20	1.221
21	1.177

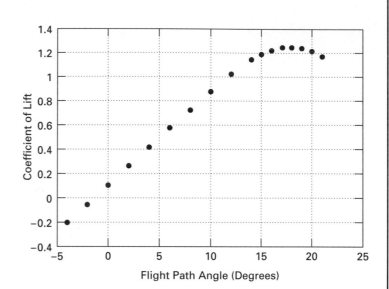

Write a program to read the data from a data file called `lift.dat`. The first line in the data file contains the number of data pairs. Each successive line in the data file contains two numbers, where the first value represents an angle (in degrees), and the second value contains C_L (the coefficient of lift determined during the wind tunnel test for the corresponding angle). Assume that the data pairs contained in the data file are in ascending order based on the angle values, with the smallest angle first and the largest angle last.

The program should then read a new angle α (alpha) from the keyboard for which we need to know the interpolated value of C_L. Using the algorithm shown earlier in this section, the program should interpolate for the lift coefficient, print the value of α and the corresponding estimate of C_L, and then prompt us for a new value of α. If we enter a value of α that is outside the range of the data, the program should print a message that the requested value is not within the interpolation region. The program terminates when we enter a value of 999 for α.

1. Problem Statement

Using linear interpolation with a wind tunnel data file, compute the lift coefficients that correspond to specified flight path angles.

2. Input/Output Description

Input—the wind tunnel data file and the values of α for which we need the interpolated values of C_L

Output—a list of the input values of α and the corresponding values of C_L

 ## 3. Hand Example

The data file that we use for testing contains the 17 data points given at the beginning of this section. Assume that we want to estimate the value of C_L that corresponds to $\alpha = 6.3$. Using this data file and the linear interpolation formula from the previous section, the estimated value of C_L is

$$C_L = 0.573 + \frac{6.3 - 6}{8 - 6} \cdot (0.727 - 0.573)$$
$$= 0.573 + 0.15(0.154)$$
$$= 0.596$$

 ## 4. Algorithm Development

Decomposition

Read data file.
Read angles and interpolate coefficients.

As we refine these steps, remember that we need to determine if the input value of α is within the interpolation region. We can do this by checking to see if α is between the first flight path angle and the last flight path angle in the data file. If the input value of α is not within this allowable interpolation region, we print a message. We also need to check for an input value of α of 999 so that we can terminate the program.

Pseudocode

```
tunnel:   read n, the number of data pairs
          read data file into arrays angle and lift
          read input angle α
          while α ≠ 999
              if angle(1) ≤ α ≤ angle(n)
                  find k where angle(k-1) ≤ α ≤ angle(k)
                  interpolate a value for C_L
                  print α, C_L
              else
                  print message that α is out of range
              read new input angle α
```

Fortran Program

```fortran
!-----------------------------------------------------------------!
program tunnel
!
!   This program reads a file of wind tunnel test data
!   and then uses linear interpolation to compute lift.
!
!       Declare and initialize variables.
        implicit none
        integer :: k, n
        real :: alpha, angle(500), C_L, lift(500)
!
!       Open file and read wind tunnel data into arrays.
        open(unit=10,file="lift.dat",status="old")
        read(10,*) n
        do k=1,n
            read(10,*) angle(k), lift(k)
        end do
!
!       Prompt user to enter angle alpha.
        print*, "Enter flight path angle in degrees:"
        read*, alpha
        do while (alpha /= 999)
!
!           Find location of alpha in the angle data.
            if ((angle(1) <= alpha).and.(alpha <= angle(n))) then
                if (alpha == angle(1)) then
                    k = 2
                else
                    k = 1
                    do while (angle(k) < alpha)
                        k = k + 1
                    end do
                end if
                C_L = lift(k-1) + &
                    (alpha - angle(k-1))/(angle(k) - angle(k-1))* &
                    (lift(k) - lift(k-1))
                print 20, alpha, C_L
   20           format(1x,"flight angle = ",f7.2/ &
                        1x,"corresponding lift coefficient = ",f7.3)
            else
                print*,"Angle out of range for this data file."
            end if
!           Prompt for next flight path angle.
            print*
            print*, "Enter flight path angle in degrees:"
            print*, "(999 to quit)"
            read*, alpha
        end do
!
!   Close file.
    close(unit=10)
!
end program tunnel
!-----------------------------------------------------------------!
```

5. Testing

We specified 500 values for arrays angle and lift, but these can be increased if we expect to need more values. Using the data file given at the beginning of this section, we tested the program with several values of α, including values outside the range of the data file. The interaction with the program is as follows:

```
Enter flight path angle in degrees:
6.3
flight angle =    6.30
corresponding lift coefficient =    0.596

Enter flight path angle in degrees:
(999 to quit)
20.5
flight angle =   20.50
corresponding lift coefficient =    1.199

Enter flight path angle in degrees:
(999 to quit)
-5
Angle out of range for this data file.

Enter flight path angle in degrees:
(999 to quit)
22.5
Angle out of range for this data file.

Enter flight path angle in degrees:
(999 to quit)
-4
flight angle =   -4.00
corresponding lift coefficient =   -0.202

Enter flight path angle in degrees:
(999 to quit)
21
flight angle =   21.00
corresponding lift coefficient =    1.177

Enter flight path angle in degrees:
(999 to quit)
10
flight angle =   10.00
corresponding lift coefficient =    0.880

Enter flight path angle in degrees:
(999 to quit)
999
```

Note that the first value computed by the program is the same value that we computed in the hand example. Also note that we tested all the special cases, including values of α that are out of range, values at the endpoints, and values at a data point.

5-3 A SELECTION SORT ALGORITHM

The topic of sorting techniques is the subject of entire textbooks and courses; therefore, this module will not attempt to present all the important aspects of sorting. Instead, we present a common sorting technique called a *selection sort.* This algorithm *sorts,* or reorders, a one-dimensional array or list into *ascending,* or low-to-high, order. With minor alterations the algorithm can be changed to one that sorts in *descending,* or high-to-low, order. In Chapter 7 we also use this sort algorithm to alphabetize character data.

The selection sort is a simple sort that is based on finding the minimum value and placing it first in the list, finding the next smallest value and placing it second in the list, and so on. We will use only one array in the algorithm. Therefore, if we need to keep the original order of the data as well as the sorted data, we should copy the original data into a second array and sort it. Since the actual number of values in the array may be less than the maximum number of values that could be stored in it, we will assume that the variable count specifies the actual number of data values to be sorted.

We begin our look at the selection sort with a hand example. Consider the following list of data values:

Original list	4.1
	7.3
	1.7
	5.2
	1.3

In this algorithm, we first find the minimum value. Scanning down the list, we find that the last value, 1.3, is the minimum. We now want to put the value 1.3 in the first position of the array, but we do not want to lose the value 4.1 that is currently in the first position. Therefore, we will exchange the values. The switch of two values requires three steps, not two as you might imagine. Consider these statements:

```
integer :: i, j
real :: x(100)
...
x(i) = x(j)
x(j) = x(i)
```

Suppose x(i) contained the value 3.0 and x(j) contained the value −1.0. The first statement will change the contents of x(i) from the value 3.0 to the value -1.0. The second statement will move the value in $x(i)$ to $x(j)$, so that both locations contain -1.0. These steps are shown below, along with the changes in the corresponding memory locations:

```
                            x(i)     x(j)
                            3.0      -1.0
            x(i) = x(j)     -1.0     -1.0
            x(j) = x(i)     -1.0     -1.0
```

You can see that these two statements do not correctly switch the values. A correct way to switch the two values is shown below, along with the changes in the corresponding memory locations:

```
                        x(i)     x(j)     hold
                        3.0      -1.0     ?
        hold = x(i)     3.0      -1.0     3.0
        x(i) = x(j)     -1.0     -1.0     3.0
        x(j) = hold     -1.0     3.0      3.0
```

Once we have switched the first value in the array with the value that has the minimum value, we search for the minimum in the values in the array starting from the second value to the last value. We then switch the second value with this minimum. We continue this until we are looking at the next-to-last and last values. If they are out of order, we switch them. At this point the entire array will be sorted into an ascending order, as shown in the following diagram:

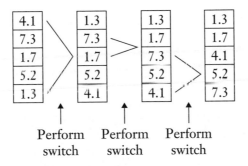

We now develop the pseudocode and Fortran statements for this selection sort.

Decomposition

Sort list of data values into ascending order.

In the following refinement notice that we do not keep track of the minimum value itself; instead, we are interested in keeping track of the subscript or location of the minimum value. We need the subscript in order to be able to switch positions with another element in the array. Since the portion of the array that we search for the next minimum gets smaller, we must keep track of the current or first position with a variable current. We

find the minimum in the positions that start with the current position and
end with the value in the last position.

Pseudocode

(Assumes that values are already stored in the array x, and that a variable
count specifies the number of valid data values in the array.)

```
selection:    current ← 1
              while current ≤ count-1
                   next_min ← current
                   k ← current+1
                   while k ≤ count
                        if x(k) < x(next_min)
                             next_min ← k
                        add 1 to k
                   switch values in x(current) and x(next_min)
                   add 1 to current
```

Fortran Statements

```
!------------------------------------------------------------------!
!
!     These statements sort the values in the real array x
!     using a selection sort. The integer count contains
!     the number of valid data values in x. The variables
!     current, k, and next_min are integers; hold is a real
!     variable.
!
!     Let current specify the current position.
      do current=1,count-1
         next_min = current
!
!        Find next minimum.
         do k=current+1,count
            if (x(k) < x(next_min)) next_min = k
         end do
!
!        Switch the current position with next minimum.
         hold = x(current)
         x(current) = x(next_min)
         x(next_min) = hold
      end do
!
!------------------------------------------------------------------!
```

Try It Try this self-test to check your memory of certain key points from Section
5-3. If you have any problems with the exercise, you should reread this
section. The solution is given at the end of this module.

Consider the following list with six elements in it:

```
31
24
63
16
21
8
```

Show the sequence of changes that occur in the list if it is sorted using the selection sort algorithm.

5-4 DYNAMIC ARRAY ALLOCATION

In FORTRAN 77 we had to define an array to the maximum possible size that we might need, even though we might actually use only a small fraction of it. In Fortran 90 we can initially define an array without specifying the size of the array. This is useful when the size of the array is computed within the program or is read from a data file.

In addition to allocating the memory needed for an array during the execution of a program, we can also release or deallocate the memory when we no longer need the array. The statements to allocate and deallocate memory are as follows:

```
allocate (array name (size))
```

```
deallocate (array name)
```

Assume that the first line of a data file contains the number of lines of valid data in the file. The following statements define arrays to store the data and allocate the memory after the number of elements has been determined:

```
integer :: count
real, dimension(:), allocatable :: x, y
...
read(9,*) count
allocate (x(count), y(count))
```

The dimension specifier on the real statement indicates that the variables are arrays, but that their sizes are unspecified at compile time. The allocatable specifier indicates that the memory is not to be assigned to these variables at compile time, but that it can be assigned during the execution of the program if an allocate statement is executed for these variables.

5-5 TWO-DIMENSIONAL ARRAYS

If we visualize a one-dimensional array as a single column of data, we can visualize a *two-dimensional array* as a group of columns, as illustrated:

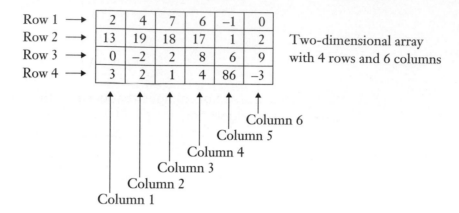

The diagram depicts an integer array with 24 elements. As in one-dimensional arrays, each of the 24 elements has the same array name. However, one subscript is not sufficient to specify an element in a two-dimensional array. For instance, if the array's name is m, it is not clear whether m(3) should be the third element in the first row or the third element in the first column. To avoid ambiguity, elements in a two-dimensional array are referenced with two subscripts: The first subscript references the row, and the second subscript references the column. Thus, m(2,3) refers to the number in the second row, third column. In our diagram, m(2,3) contains the value 18.

Array Storage

Two-dimensional arrays must be specified with a type statement or a dimension statement, but not both. The following type statements reserve storage for a two-dimensional integer array j with seven rows and four columns, a one-dimensional real array b of ten elements, a two-dimensional real array c with three rows and five columns, and a two-dimensional real array num with five rows and two columns:

```
integer :: j(7,4)
real :: b(10), c(3,5), num(5,2)
```

This next statement reserves storage for a two-dimensional array with three rows and three columns:

```
real :: r(0:2,-1:1)
```

The first element in the first row is referenced as r(0,-1), the first element in the second row is referenced as r(1,-1), and so on.

Initialization and Computations

Two-dimensional arrays can be initialized with assignment statements, with input statements, and with the data statement. A two-dimensional array name can be used without subscripts in assignment statements, and the specifed operations are performed element-by-element, as illustrated in the following statements:

```
real :: b(2,3), c(2,3)
...
c = b + c
```

These statements define two arrays, each with two rows and three columns. The assignment statement specifies that the values in arrays b and c are to be added in an element-by-element way and then stored in c. Thus, these statements are equivalent to

```
integer :: i, j
real :: b(2,3), c(2,3)
...
do i=1,2
   do j=1,3
      c(i,j) = b(i,j) + c(i,j)
   end do
end do
```

If an array name is used in input statements, data statements, and output statements, the array is accessed in column order. In this module we will explicitly use subscripts with two-dimensional arrays in order to be clear whether we are referencing the array elements in row order or in column order. (Row order is an order that goes across the rows, as in c(1,1), c(1,2), c(1,3), c(2,1), ... ; column order is an order that goes down the columns, as in c(1,1), c(2,1), c(1,2), ...) It is common notation to use i for the row subscript and j for the column subscript.

EXAMPLE 5-3

Array Initialization, area

Define an array area with five rows and four columns. Fill it with the values shown:

1.0	1.0	2.0	2.0
1.0	1.0	2.0	2.0
1.0	1.0	2.0	2.0
1.0	1.0	2.0	2.0
1.0	1.0	2.0	2.0

SOLUTION 1

```
integer :: i
real :: area(5,4)
...
do i=1,5
   area(i,1) = 1.0
   area(i,2) = 1.0
   area(i,3) = 2.0
   area(i,4) = 2.0
end do
```

SOLUTION 2

```
integer :: i, j
real :: area(5,4)
data ((area(i,j),i=1,5),j=1,4) /10*1.0,10*2.0/
```

. . . .

EXAMPLE 5-4 **Array Initialization, sum**

Define and fill the integer array sum as shown:

1	1	1
2	2	2
3	3	3
4	4	4

SOLUTION 1

If we observe that each element of the array contains its corresponding row number, then we can use the following solution:

```
integer :: i, j, sum(4,3)
...
do i=1,4
   do j=1,3
      sum(i,j) = i
   end do
end do
```

SOLUTION 2

The following data statement also initializes the array correctly:

```
integer :: i, j, sum(4,3)
data ((sum(i,j),j=1,3),i=1,4) /3*1,3*2,3*3,3*4/
```

. . . .

EXAMPLE 5-5 **Identity Matrix**

A matrix is a structure used to store data that can be represented as a rectangular grid of numbers. Therefore, a matrix looks very similar to a two-dimensional array, except that a matrix has brackets on the sides while a two-dimensional array is shown inside a grid. When a matrix is used in a Fortran program, we store it in a two-dimensional array. When using matrix operations to solve engineering and science problems, we frequently use a matrix called an *identity matrix*. This matrix has the same number of rows as columns, so it is also called a *square matrix*. The identity matrix contains all 0s except for the main diagonal elements, which are 1s. (The main diagonal is composed of elements that have the same value for row number and for column number.) An identity matrix with five rows and five columns is shown on the next page.

$$\begin{bmatrix} 1 & 0 & 0 & 0 & 0 \\ 0 & 1 & 0 & 0 & 0 \\ 0 & 0 & 1 & 0 & 0 \\ 0 & 0 & 0 & 1 & 0 \\ 0 & 0 & 0 & 0 & 1 \end{bmatrix}$$

Define and fill a real array with these values.

SOLUTION

Because the value 1 appears at different positions in each row of the array, we cannot use the same type of solution we used in Example 5-4. If we list the elements that contain the value 1.0, we find that they are positions (1,1), (2,2), (3,3), (4,4), and (5,5); thus, the row and the column number are the same value. Recognizing this pattern, we can initialize all elements in the array to zero, and then change the values on the main diagonal to the value 1:

```
integer :: i, j
real :: identity(5,5)=0.0
...
do i=1,5
    identity(i,i) = 1.0
end do
```

Input and Output

The main difference between using values from a one-dimensional array and using values from a two-dimensional array is that the latter requires two subscripts. Most loops used in the reading or printing of two-dimensional arrays are therefore nested loops. You can also use implied loops with two-dimensional array I/O. Avoid using only the array name on I/O statements with two-dimensional arrays; using the array name with no subscripts is equivalent to listing all the values in the array but in an order that goes down the columns instead of across the rows.

EXAMPLE 5-6

Reading Array Values from a Data File

Suppose that a data file contains temperature measurements taken at five times during an experiment. For each time, temperatures are taken at three locations in the jet engine being tested and stored on the same line in a data file. Read this data and store it in a real array temperature that has been dimensioned with five rows and three columns. Assume that the data file has already been opened.

CORRECT SOLUTION 1

```
integer :: i
real :: temperature(5,3)
...
do i=1,5
    read(10,*) temperature(i,1),temperature(i,2), &
               temperature(i,3)
end do
```

This solution uses a single do loop and lists the variable names for each row on the read statement.

CORRECT SOLUTION 2

```
integer :: i, j
real :: temperature(5,3)
...
do i=1,5
   read(10,*) (temperature(i,j),j=1,3)
end do
```

This solution executes exactly like solution 1 because the implied do loop on the read statement is equivalent to listing the variables.

INCORRECT SOLUTION 1

```
integer :: i, j
real :: temperature(5,3)
...
do i=1,5
   do j=1,3
      read(10,*) temperature(i,j)
   end do
end do
```

This solution is the equivalent of 15 read statements. Since each read statement reads a new line in the data file, this solution tries to read each value from a new data line. Thus, the first number on line 1 will be read into temperature(1,1), the first number on line 2 will be read into temperature(1,2), and so on. The values are being stored in the wrong locations, and an execution error will occur when the program tries to read past the end of the data file.

INCORRECT SOLUTION 2

```
real :: temperature(5,3)
...
read(10,*) temperature
```

This solution is equivalent to one read statement with all 15 variables written on it, but the variables are in column order, as in temperature(1,1), temperature(2,1), temperature(3,1), temperature(4,1), temperature(5,1), temperature(1,2), and so on. Since the values in the data file are in row order, not column order, the values will be read into the wrong locations. Unfortunately, no error is detected by the computer, so we may not detect this error unless we carefully test our program.

. . . .

EXAMPLE 5-7   **Print Array Values**

In this example we assume that we have correctly read the temperature data from the data file described in Example 5-6. We now want to print the data, with three values per line.

CORRECT SOLUTION 1

```
integer :: i
real :: temperature(5,3)
...
do i=1,5
    print*, temperature(i,1), temperature(i,2), &
            temperature(i,3)
end do
```

This solution uses a single do loop and lists the variable names for each row on the print statement. Five lines of output will be printed, with three values per line, using a list-directed format.

CORRECT SOLUTION 2

```
integer :: i
real :: temperature(5,3)
...
do i=1,5
    print 10, temperature(i,1), temperature(i,2), &
            temperature(i,3)
10    format(1x,3(f5.2,2x))
end do
```

This solution also prints three values per line using the specified format instead of a list-directed format.

INCORRECT SOLUTION

```
integer :: i
real :: temperature(5,3)
...
do i=1,5
    print 10, temperature(i,1), temperature(i,2), &
            temperature(i,3)
10    format(1x,f5.2)
end do
```

This solution prints one value per line instead of three because the format contains only one specification. When we run out of format specifications, we print the current buffer and back up in the format list to get a specification.

. . . .

Try It

Try this self-test to check your memory of some key points from Section 5-5. If you have any problems with the exercises, you should reread this section. The solutions are given at the end of this module.

Problems 1–3 contain statements that initialize and print two-dimensional arrays. Draw the array and indicate the contents of each position in the array. Then show the output from each set of statements. Assume that each set of statements is independent of the others.

```
1.      integer :: check(5,4), i, j
        ...
        do i=1,5
           do j=1,4
              check(i,j) = 2*(i + j)
           end do
        end do
        print 15, (check(5,j),j=1,4)
     15 format(1x,4i5)
2.      integer :: i, j, k(3,3)
        ...
        do i=1,3
           k(i,1) = i
           k(i,2) = k(i,1) + 1
           k(i,3) = k(i,1) - 1
        end do
        do i=1,3
           print 20, k(i,1), k(i,2), k(i,3)
     20    format(1x,3i4)
        end do
3.      integer :: i, j
        real :: distance(2,3), sum=10.0
        ...
        do j=1,3
           do i=1,2
              sum = sum + 1.5
              distance(i,j) = sum
           end do
        end do
        do i=1,2
           print 15, (distance(i,j),j=1,3)
     15    format(1x,3f5.1)
        end do
```

Intrinsic Functions

The intrinsic functions in Table 5-1 that were discussed in the section on one-dimensional arrays can also be used with two-dimensional arrays. For example, if the array b is defined with two rows and three columns, the following statements will print the sum of the elements in the array and the product of the elements in the array:

```
real :: b(2,3)
...
print*, "The sum of the array elements is ", sum(b)
print*, "The product of the array elements is ", &
        product(b)
```

Of course, these statements assume that values have been given to all the array elements in b before the print statements are executed.

5-6 Application   **POWER PLANT DATA ANALYSIS**

Power Engineering

The following table of data represents typical power outputs in megawatts from a power plant over a period of eight weeks. Each row represents one week's data; each column represents data taken on the same day of the week. The data is stored one row per data line in a data file called plant.dat.

	Day 1	Day 2	Day 3	Day 4	Day 5	Day 6	Day 7
Week 1	207	301	222	302	22	167	125
Week 2	367	60	120	111	301	499	434
Week 3	211	62	441	192	21	293	316
Week 4	401	340	161	297	441	117	206
Week 5	448	111	370	220	264	444	207
Week 6	21	313	204	222	446	401	337
Week 7	213	208	444	321	320	335	313
Week 8	162	137	265	44	370	315	322

A program is needed to read the data, analyze it, and print the results in the following composite report:

```
                 Composite Information

Average Daily Power Output = xxx.x megawatts

Number of Days with Greater than Average Power Output = xx

Day(s) with Minimum Power Output:
          Week x     Day x
          . . .
```

1. Problem Statement

Analyze a set of data from a power plant to determine its average daily power output, the number of days with greater-than-average output, and the day (or days) that had minimum power output.

2. Input/Output Description

Input—8 weeks of daily power output stored in a data file

Output—a report summarizing the power output for the 8 weeks

3. Hand Example

For the hand example we use a smaller set of data, but one that still maintains the two-dimensional array form. Consider this set of data:

	Day 1	Day 2
Week 1	311	405
Week 2	210	264
Week 3	361	210

First, we must sum all the values and divide by 6 to determine the average, which yields 1761/6, or 293.5 megawatts. Second, we compare each value to the average to determine how many values were greater than the average. In this small set of data, three values were greater than the average. Third, we must determine the number of days with minimum power output, which involves two steps: going through the data again to determine the minimum value, and then going back through the data to find the day or days with the minimum power value. Then we can print its/their position(s) in the array. Using the small set of data, we find that the minimum value is 210 and that it occurred on two days. Thus, the output from our hand example is

```
               Composite Information

Average Daily Power Output = 293.5 megawatts

Number of Days with Greater than Average Power Output = 3

Day(s) with Minimum Power Output:
        Week 2    Day 1
        Week 3    Day 2
```

4. Algorithm Development

Before we decompose the problem solution, it is important to spend some time considering the best way to store the data that we need for the program. Unfortunately, once we become comfortable with arrays, we tend to overuse them. Using an array complicates our programs because of the subscript handling. We should always ask ourselves, "should we really use an array for this data?" If the individual data values will be needed more than once, an array is probably required. An array is also necessary if the data is not in the order needed. In general, arrays are helpful when we must read all the data before we can go back and begin processing it. However, if an average of a group of data values is all that is to be computed, we probably do not need an array; as we read the values, we can add them to a total and read the next value into the same memory location as the previous value. The individual values are not needed again because all the information required is now in the total.

Now let us look at our specific problem and determine whether or not we need to use an array. First we need to compute an average daily power output. Then we need to count the number of days with output greater than average, which requires that we compare each output value to the average. For this application we need to store all the data in an array, and a two-dimensional array is the best choice of data structure.

We would like to minimize the number of passes through the data. We can compute the sum of the data points in one loop. However, we must make a separate pass through the data to determine how many values are greater than the average. Because the number of days with greater-than-average output is printed before we print the specific day or days that has/have minimum output, we need separate loops for these operations.

Decomposition

Read data.
Compute information.
Print information.

Pseudocode

power_plant: read data
                      compute and print average power
                      count days with above-average power
                      print count
                      determine minimum power
                      print day and week number for days
                              with minimum power

Fortran Program

```
!-------------------------------------------------------------!
program power_plant
!
!   This program computes and prints a composite report
!   summarizing eight weeks of power plant data.
!
!
!      Declare and initialize variables.
       implicit none
       integer :: count=0, i, j, minimum, power(8,7)
       real :: average
!
!      Open file and read data into the array.
       open(unit=12,file="plant.dat",status="old")
       do i=1,8
          read(12,*) (power(i,j),j=1,7)
       end do
!
!      Compute average power and determine number
!      of days with power greater than the average.
       average = sum(power)/56.0
       do i=1,8
          do j=1,7
             if (power(i,j) > average) count = count + 1
          end do
       end do
```

```
!
!       Print initial part of the report.
        print 30, average, count
   30 format(1x,15x,"Composite Information"// &
           1x,"Average Daily Power Output = ", &
           f5.1," megawatts"// &
           1x,"Number of Days with Greater Than ", &
           "Average Power Output = ",i2/)
!
!       Determine and print days with minimum power output.
        minimum = minval(power)
        print 35
   35 format(1x,"Day(s) with Minimum Power Output:")
        do i=1,8
           do j=1,7
              if (power(i,j) == minimum) print 40, i, j
   40         format(1x,9x,"Week ",i1,"    Day ",i1)
           end do
        end do
!
!       Close file.
        close(unit=12)
!
end program power_plant
!-----------------------------------------------------------------!
```

5. Testing

This program should be tested in stages; again, the decomposition gives a good idea of the overall steps involved and can thus be used to identify the steps that should be tested individually. Remember that one of the most useful tools for debugging is the print statement; use it to print the values of key variables in loops that may contain errors.

The output from this program, using the data file given at the beginning of this section, is

```
            Composite Information

Average Daily Power Output = 259.2 megawatts

Number of Days with Greater Than Average Power Output = 30

Day(s) with Minimum Power Output:
        Week 3    Day 5
        Week 6    Day 1
```

5-7 MULTIDIMENSIONAL ARRAYS

Fortran allows *multidimensional arrays* with as many as seven dimensions. We can easily visualize a three-dimensional array such as a cube. We are also familiar with using three coordinates, x, y, and z, to locate points. This idea extends into subscripts. The following three-dimensional array could be defined with this statement:

```
real :: t(3,4,4)
```

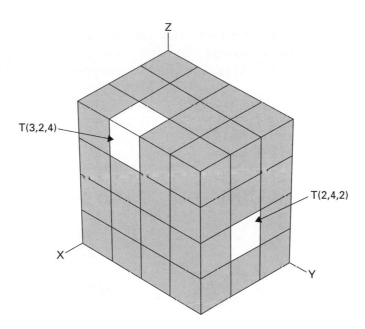

If we use the three-dimensional array name without subscripts, we access the array with the first subscript changing fastest, the second subscript changing second fastest, and the third subscript changing slowest. Thus, using the array t from the previous diagram, these two statements are equivalent:

```
read*, t
read*, (((t(i,j,k),i=1,3),j=1,4),k=1,4)
```

It should be evident that three levels of nesting in do loops are often needed to access a three-dimensional array.

Most applications do not use arrays with more than three dimensions, probably because visualizing more than three dimensions seems too abstract. However, we now present a simple scheme that may help you to picture even a seven-dimensional array.

A four-dimensional array can be visualized as a row of three-dimensional arrays. The first subscript specifies a unique three-dimensional array. The other three subscripts specify a unique position in that array.

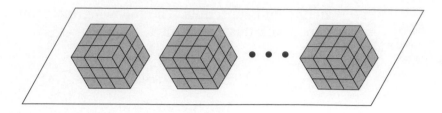

A five-dimensional array can be visualized as a block or grid of three-dimensional arrays. The first two subscripts specify a unique three-dimensional array. The other three subscripts specify a unique position in that array.

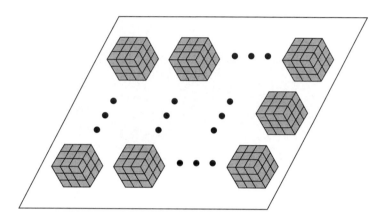

A six-dimensional array can be visualized as a row of blocks or grids. One subscript specifies the grid. The other five subscripts specify the unique position in the grid.

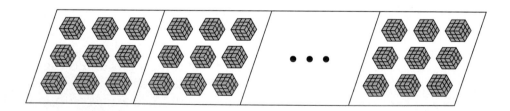

A seven-dimensional array can be visualized as a grid of grids or a grid of blocks. Two subscripts specify the grid. The other five subscripts specify the unique position in the grid.

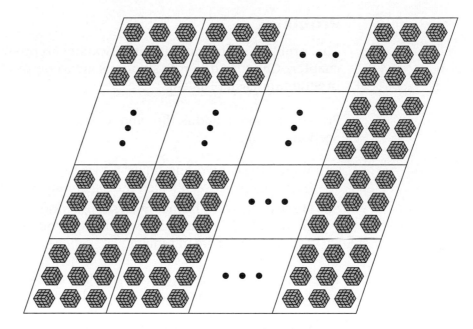

Now that you can visualize multidimensional arrays, a natural question is, "What dimension array do I use for solving a problem?" There is no single answer. A problem that can be solved with a two-dimensional array of four rows and three columns can also be solved with four one-dimensional arrays of three elements each. Usually the data fit one array form better than another; you should choose the form that is the easiest for you to work with in your program. For example, if you have census data from 10 countries over the period 1950–1980, you would probably use an array with 10 rows and 31 columns, or 31 rows and 10 columns. If the data represent the populations of 5 cities from each of 10 countries for the period 1950–1980, a three-dimensional array would be most appropriate; the three subscripts would represent year, country, and city.

SUMMARY

In this chapter you learned how to use an array—a group of storage locations that all have a common name but are distinguished by one or more subscripts. Arrays are one of the most powerful elements in Fortran because they allow you to keep large amounts of data easily accessible to your programs. The remaining chapters in this module rely heavily on arrays for storing and manipulating data. This chapter also presented a selection sort algorithm.

Key Words

array	one-dimensional array
ascending order	selection sort
descending order	sort
element	square matrix
identity matrix	subscript
implied do loop	two-dimensional array
multidimensional array	

Problems

This problem set begins with modifications to programs developed in this chapter. Give the decomposition, pseudocode or flowchart, and Fortran program for each problem.

Problems 1 and 2 modify the wind tunnel analysis program `tunnel` given in Section 5-2 on page 132.

1. Modify the wind tunnel analysis program so that it prints the two flight path angles between which the maximum coefficient of lift occurs.
2. Modify the wind tunnel analysis program so that it asks the user to enter the flight path angle in degrees, but assume that the data file contains the angle in radians.

Problem 3 modifies the power plant data analysis program `power_plant` given in Section 5-6 on page 147.

3. Modify the power plant data analysis program so that it reads a value n that determines the number of weeks that will be used for the report. Assume that n will never be more than 8.

For problems 4–6 assume that k, a one-dimensional array of 50 integer values, has already been filled with data.

4. Give Fortran statements to find and print the minimum value of k and its position or positions in the array in the following form:

    ```
    minimum value of k is
    k(xx) = xxxxx
    ```

5. Give Fortran statements to count the number of positive values, zero values, and negative values in k. The output form should be

    ```
    xxx positive values
    xxx zero values
    xxx negative values
    ```

6. Give Fortran statements to replace each value of k with its absolute value, and then print the array k with two values per line.

Develop the following programs and program segments. Use the five-step problem-solving process for all complete programs.

7. Give Fortran statements to print the last 10 elements of a real array m of size n. For instance, if m contains 25 elements, the output form is

    ```
    m( 16) = xxx.x
    m( 17) = xxx.x
    ...
    m( 25) = xxx.x
    ```

8. Give Fortran statements to interchange the first and one-hundredth elements, the second and ninety-ninth elements, and so on of the integer array numbers that contains 100 integer values. See the diagram that follows:

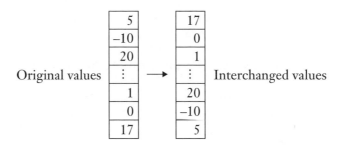

Original values → Interchanged values

(*Hint:* You will need a temporary variable when you switch values.)

9. When a plot is made from experimental data, sometimes the scatter of the data points is such that it is difficult to select a "best representative line" for the plot. In such a case the data can be adjusted to reduce the scatter by using a "moving average" mathematical method of finding the average of three points in succession and replacing the middle value with this average.

    Write a complete program to read an array y of 20 real values from a file lab1.dat, where the values are entered one per line. Build an array z of 20 values, where z is the array of adjusted values. That is, z(2) is the average of y(1), y(2), and y(3); z(3) is the average of y(2), y(3), and y(4); and so on. Notice that the first and last values of y cannot be adjusted and should be moved to z without being changed. Do not destroy the original values in y. Print the original and the adjusted values next to each other in a table.

10. A truck leasing company owns 12 delivery vans that are leased to several operators. Maintenance hours for each truck are allocated using a maintenance rate multiplied by the total number of hours accumulated by the entire fleet. The maintenance rate for each truck is determined from its percentage of the total fleet hours and the following table:

Percent of Hours	Maintenance Rate
0.00–9.99	0.02
10.00–24.99	0.04
25.00–100.00	0.06

A vehicle identification number and the monthly hours of use are entered in a data file hours.dat. Write a complete program to read this integer data, convert each truck's hours to a percentage of the total fleet hours, and compute its maintenance hours using the maintenance rate. Round the calculated maintenance hours to the nearest hour, and ensure that each truck is allotted a minimum of one hour of maintenance each month. Print the following report:

```
Monthly Maintenance Report
ID      Hours   Percent   Maintenance Hours
xxx     xxxx    xxx.x     xxx
...
xxx     xxxx    xxx.x     xxx
Totals  xxxx    xxx.x     xxx
```

Test your program with the following input data:

ID	Hours	ID	Hours
002	61	037	66
009	83	040	70
012	101	043	122
016	55	044	136
025	410	045	23
036	97	046	142

11. Assume that the reservations for an airplane flight have been stored in a file called `flight.dat`. The plane contains 38 rows, with 6 seats in each row. The seats in each row are numbered 1 through 6 as follows:

Code	Location
1	Window seat, left side
2	Center seat, left side
3	Aisle seat, left side
4	Aisle seat, right side
5	Center seat, right side
6	Window seat, right side

The file `flight.dat` contains 38 lines of information corresponding to the 38 rows. Each line contains 6 values corresponding to the 6 seats. The value for any seat is either 0 or 1, representing either an empty or an occupied seat.

Write a complete program to read the flight information into a two-dimensional array called `seat`. Find and print all pairs of adjacent seats that are empty. Adjacent aisle seats should not be printed. If all three seats on one side of the plane are empty, then two pairs of adjacent seats should be printed. Print this information in the following manner:

```
Available Seat Pairs
Row  Seats
xx   x,x
...
```

If no pairs of seats are available, print an appropriate message.

6 External Procedures

Oil Well Production Computer programs are used to analyze many types of information. In a large oil-producing company, they would be used to keep track of oil well production and to develop mathematical models for predicting oil well production. Computers would also be used to perform statistical analysis on the data to determine averages and trends in the data over time. Computers are also used to analyze results from experiments that collect data from seismometers (sensors that detect earth motion) and then attempt to model the geological structure under the ground. This information (often transmitted by satellite) is not only useful in predicting regions likely to contain oil, but it is also useful to engineers and scientists who are attempting to predict accurately the size and location of major earthquakes.

INTRODUCTION

As our programs become longer and more complicated, we find it harder to maintain program readability and simplicity. We also find that we frequently need to perform the same set of operations at more than one location in our programs. We can solve these problems by using *procedures,* which are groups of statements that are defined separately and then referenced when we need them in our programs. Fortran has two main types of procedures: functions and subroutines. This chapter begins with an overview of programming with procedures. We then review intrinsic functions, such as the square root and logarithm functions, which are in the compiler and accessible from your program. The chapter then focuses on *external procedures*—user-defined functions and subroutines that are completely separate from the main program and from each other. (Internal procedures are discussed in Sections 8-2 and 8-3.) We learn how to write our own functions to perform computations unique to our applications. Such functions are quite useful in solving engineering and science problems because many of our solutions involve arithmetic computations. Next we concentrate on *subroutines,* which are procedures that allow us to return many values as opposed to a single function value. Subroutines can also be used to read data files and print reports. We will use both functions and subroutines frequently in the remainder of this module to illustrate their importance in making programs simpler and more readable.

6-1 PROGRAMMING WITH PROCEDURES

In previous chapters we stressed the importance of using while loops, do loops, and if structures as essential ingredients in writing structured programs. Another key element in structuring program logic is the use of procedures that allow us to write programs composed of nearly independent segments or routines. Some procedures are part of the language, such as intrinsic functions, and some procedures are user defined. In Fortran user-defined procedures are implemented as functions or subroutines. Functions and subroutines look very much like programs except that they begin with a `function` statement or a `subroutine` statement instead of a `program` statement.

When we decompose a problem solution into a series of sequentially executed steps, we are decomposing the problem into steps that probably could be easily structured into functions and subroutines. The following are some important advantages to breaking programs into procedures:

- You can write and test each procedure separately from the rest of the program.
- Debugging is easier because you are working with smaller sections of the program.
- Procedures can be used in other programs without rewriting or retesting.
- Programs are more readable and thus more easily understood because of the procedure structure.
- Several programmers can work on different procedures of a large program relatively independent of one another.
- Individual parts of the program become shorter and therefore simpler.
- A procedure can be used several times by the same program.

The decomposition diagram shows the sequential steps necessary to solve a problem. Another type of diagram, the *structure chart,* is also very useful as we decompose our problem solution into smaller problems. While the decomposition diagram outlines the sequential operations needed, the structure chart outlines the procedures but does not indicate the order in which they are to be executed.

The following diagram contains a structure chart for a program that we will develop in Section 6-4. In the program we read oil well production data from a data file. For each oil well, we read the daily production in barrels, compute a daily average for the week, and print this information in the report. In addition, we keep summary information for all the wells and print it at the end of the report. A function is used to count the number of days that the production falls below the average production for each oil well during each week.

It is important to distinguish between a decomposition diagram and a structure chart. A decomposition diagram shows the sequential order of the steps of a solution; it does not identify any of the steps as procedures. A structure chart, on the other hand, clearly defines the procedure definitions for a program but does not show the order in which the procedures are used. Thus, both charts are useful in describing the algorithm that is being developed for a problem solution because they represent different information about the solution.

6-2   INTRINSIC FUNCTIONS

A *function* computes a single value, such as the square root of a number or the average of an array. You have already used functions in the form of intrinsic functions, such as sqrt and sin. These intrinsic functions are in the compiler and are accessible directly from your program. Many of the intrinsic functions (or library functions) available in Fortran 90 are listed in Appendix A. You should read through the list so that you are aware of the types of operations that can be performed with intrinsic functions. When you need to use one of these operations, you can refer to Appendix A for details on how to use that specific function. Chapter 2 introduced intrinsic functions, and the following list summarizes their main components:

- The function name and its input values (or arguments) collectively represent a single value.
- An intrinsic function can never be used on the left side of an equal sign in an assignment statement.

- The name of an intrinsic function implicitly determines the type of output from the function, unless it is a generic function.
- The arguments of an intrinsic function are generally of the same type as the function itself. For a few exceptions, refer to the list of intrinsic functions in Appendix A.
- The arguments of an intrinsic function must be enclosed in parentheses.
- The arguments of an intrinsic function may be constants, variables, expressions, or other functions.

Generic functions accept arguments of any allowable type and return a value of the same type as the argument. Thus, the generic function abs returns an integer absolute value if its argument is an integer, but it returns a real absolute value if its argument is real. The table in Appendix A identifies generic functions.

6-3   FUNCTION PROCEDURES

Engineering and science applications often require a computation that is not included in the intrinsic function list. If the computation is needed frequently or requires several steps, we should implement it as a user-written function instead of listing all the computations each time we need them.

A function procedure, which is a program itself, is separate from the *main program* which begins with the program statement. A function begins with a nonexecutable statement that identifies the type of value that the function will return, assigns the function a name, and specifies an argument list, as shown in the following general form:

> *function-type* function *function-name(argument list)*

Because a function is separate from the main program, it must end with an end statement. The function should also contain a return statement, which returns control to the statement that referenced the function. The general form of the return statement is

> return

The rules for choosing a function name are the same as those for choosing a program name. The following statements illustrate a simple example with a structure chart, a main program, and a function procedure that computes the average of three values. The function is named average__1 instead of average because we are going to develop several functions to compute averages in this section.

```
!-----------------------------------------------------------------!
program main_1
!
!   This is a sample main program that reads three
!   values and uses a function to compute their average.
!
!
!     Define variables.
      implicit none
      real :: average_1, x, y, z
!
!     Prompt user to enter values.
      print*,"Enter three values:"
      read*, x, y, z
!
!     Compute and print the average of the values.
      print 5, average_1(x,y,z)
    5 format(1x,"average = ",f5.2)
!
end program main_1
!-----------------------------------------------------------------!
real function average_1(x1, x2, x3)
!
!   This function computes the average of 3 values.
!
!
!     Declare variables.
      implicit none
      real :: x1, x2, x3
      intent (in) :: x1, x2, x3
!
!     Compute average.
      average_1 = (x1 + x2 + x3)/3.0
!
!     Function return.
      return
!
end function average_1
!-----------------------------------------------------------------!
```

Note that the declaration of the function arguments within the function included an additional statement that specifies the intent of the argument:

```
real :: x1, x2, x3
intent (in) :: x1, x2, x3
```

The `intent` statement indicates that these arguments represent information that is transferred to the function, and thus the value of these arguments should not be changed in the function. This characteristic or attribute could also have been specified on the type statement:

```
real, intent (in) :: x1, x2, x3
```

It is possible for arguments to be specified with an intent of `out` (which means that the argument may be given a value by the function, which is then returned to the main program) or `inout` (which means that the argument will bring information to the function and also return information to the main program). In general, if we need to return more than one value to the main program, we will use a subroutine procedure. The functions that we develop will return only one value, and that value will be returned through the function name.

Several rules must be observed in writing a function procedure. In these rules we assume that the function is referenced from a main program; however, a function can also be referenced from another function or another subroutine, and the rules are then modified accordingly.

- The function arguments referenced in the main program are the *actual arguments.* They should match in type, number, and order the *dummy arguments* used in the `function` statement. In the preceding example the actual arguments are x, y, z; the dummy arguments are x1, x2, x3. The actual argument x corresponds to x1, y corresponds to x2, and z corresponds to x3.
- If one of the arguments is an array, its dimensions must be specified in both the main program and the function.
- The value to be returned to the main program is stored in the function name using an assignment statement.
- When the function is ready to return control to the statement in the main program that referenced it, a `return` statement is executed. A function may contain more than one `return` statement.
- A function can contain references to other functions.
- A function is usually placed immediately after the main program, but it also may appear before the main program. If you have more than one function, the order of the functions does not matter as long as each function is completely separate from the other functions.
- A main program and its procedures can be stored in the same program file or in separate files. If they are stored in separate files, it is necessary to link them before the main program can be executed. The statements required to perform the linking depend on the compiler and the operating system.
- The same statement numbers can be used in both a function and the main program. No confusion occurs as to which statement is referenced because the function and the main program are completely separate. Similarly, a function and a main program can use the same variable name, such as `sum`, to store different sums as long as the variable `sum` is not an argument of the function.

- Since a function returns a value, the function name should appear in a type statement in the main program.
- An intent of in is specified for all dummy arguments. If a procedure needs to return more than one value, a subroutine is used instead of a function.
- If a variable is initialized on a type statement within a procedure, the value is initialized each time the procedure is referenced.
- In a flowchart the following special symbol is used to show that the operations indicated are performed in a function:

The following series of examples illustrates the use of these rules in writing and using user-defined functions. Because a function subprogram is a program separate from the main program, you should approach function design and implementation much as you would a complete program. Follow the same guidelines for developing a decomposition diagram, and then refine it using pseudocode or flowcharts until it is detailed enough to convert into Fortran. Because the function receives values through the argument list, the pseudocode or flowchart should begin with the name of the function and the variables that are the dummy arguments to the function.

As stated earlier, when arrays are used as arguments in a function, they must be dimensioned in the function subprogram as well as in the main program. Generally, the array should have the same size in the function as it does in the main program; however, when the size of an array is an argument to the subprogram, we use a technique called *variable dimensioning*. This technique allows us to specify an array of variable size in the subprogram. The argument value then sets the size of the array when the subprogram is executed. Example 6-1 illustrates the use of an array with a fixed size, and Example 6-2 illustrates the use of an array with a variable size. (Note that variable dimensioning refers to dimensioning of an array used as a dummy argument in a subprogram. Any array that is not a dummy argument but is defined and used in a subprogram must be dimensioned in the subprogram with a constant because it is not dimensioned in the main program.)

EXAMPLE 6-1   ## Array Average, Fixed Array Size

Write a function that receives an array of 20 real values. Compute the average of the array and return it as the function value.

SOLUTION

Step 1 is to state the problem clearly:

> Write a function that computes the average of a real array.

Step 2 is to describe the input and output:

Input—array of 20 real values
Output—array average

Step 3 is to work a simple example by hand. Let the input array contain the following values:

$$-2, 36, 24, -3, 19, 21, 8, 2, 5, 38, 16, -4, 2, 7, 17, 24, 9, -3, 6, 0$$

The average is then $\dfrac{222}{20} = 11.1$.

Step 4 is to develop an algorithm, starting with the decomposition.

Decomposition

Compute average of array x.
Return.

Pseudocode

```
average_2(x):    sum ← 0.0
                 k ← 1
                 while k ≤ 20
                         add x(k) to sum
                         add 1 to k
                 average ← sum/20.0
                 return
```

Fortran Function

```
!------------------------------------------------------------------!
real function average_2(x)
!
!   This function computes the average of a
!   real array containing 20 values.
!
!
!    Declare variables.
     implicit none
     integer :: k
     real :: sum=0, x(20)
     intent (in) :: x
!
!    Compute average.
     do k=1,20
        sum = sum + x(k)
     end do
     average_2 = sum/20.0
```

```
!
!     Function return.
      return
!
end function average_2
!------------------------------------------------------------------!
```

Step 5 is to test the program. A portion of a main program that might use this function is

```
!     Declare variables
      implicit none
      real :: average_2, quiz_average, scores(20)
!
!     Prompt the user to enter the quiz scores.
      print*, "Enter 20 quiz scores:"
      read*, scores
!
!     Compute and print quiz average.
      quiz_average = average_2(scores)
      print*, "quiz average is ",quiz_average
      ...
```

. . . .

EXAMPLE 6-2

Median Value, Variable Array Size

The median of a list of sorted numbers is defined as the number in the middle. If the list has an even number of values, the median is defined as the average of the two middle values. Write a function called median that has two arguments: a real array and an integer that specifies the number of values in the array. The function should assume that the elements of the array have already been sorted. Return the median value of the array as the function value.

SOLUTION

Step 1 is to state the problem clearly:

Write a function to determine the median value of an array.

Step 2 is to describe the input and output:

Input—a real array and an integer that specifies the number of values in the array
Output—median of the array

Step 3 is to work a simple example by hand. In the list − 5, 2, 7, 36, and 42, the median is the number 7. In the list − 5, 2, 7, 36, 42, and 82, the median is (7 + 36)/2, or 21.5.
Step 4 is to develop an algorithm, starting with the decomposition.

Decomposition

Determine median of array x.
Return.

We can use the mod function to determine if the number of elements (represented by the variable n) in the array is odd or even. If n is odd, we must decide which subscript refers to the middle value. For example, if n = 5, we want the median to refer to the third value, which is referenced by $(n/2) + 1$. Recall that we are dividing two integers, and the result will be truncated to another integer. If n is even, we want to refer to the two middle values and compute their average. If n = 6, we want to use the third and fourth values, which can be referenced by $(n/2)$ and $(n/2) + 1$.

Pseudocode

median(x,n):   if n is odd, then

$$\text{median} \leftarrow x\left(\frac{n}{2} + 1\right)$$

else

$$\text{median} \leftarrow \frac{x\left(\frac{n}{2}\right) + x\left(\frac{n}{2} + 1\right)}{2}$$

return

As we convert the pseudocode into Fortran, we must remember to specify that the function median is a real function. (The default will be an integer.)

Fortran Function

```
!-------------------------------------------------------------!
real function median(x,n)
!
!   This function determines the median value
!   in a sorted array of real numbers.
!
!
!     Declare variables.
      implicit none
      integer, :: n
      real, :: x(n)
      intent (in) :: x, n
!
!     Determine median.
      if (mod(n,2) /= 0) then
         median = x(n/2 + 1)
      else
         median = (x(n/2) + x(n/2 + 1))/2.0
      end if
!
!     Function return.
      return
!
end function median
!-------------------------------------------------------------!
```

Step 5 is to test the program. This function could be tested by a program with the following structure chart and statements:

Fortran Program

```
!-------------------------------------------------------------!
program main_2
!
!  This is a sample main program that reads sorted values
!  and then uses a function to determine their median.
!
!
!    Define variables.
     implicit none
     integer :: k, n
     real :: median, x(10)
!
!    Prompt user to enter values.
     print*, "Enter number of values in the list (<11)"
     read*, n
     print*, "Now enter the values in ascending order:"
     read*, (x(k), k=1,n)
!
!    Print the median of the values.
     print 5, median(x,n)
   5 format(1x,"median = ",f5.2)
!
end program main_2
!-------------------------------------------------------------!
                    (median function goes here)
!-------------------------------------------------------------!
```

. . . .

EXAMPLE 6-3   **Count of Negative Values in a Two-Dimensional Array**

Write a function that will receive an array of integers with five rows and seven columns. The function should return a count of negative values in the array.

SOLUTION

Step 1 is to state the problem clearly:

Write a function that determines the number of negative values in a two-dimensional array.

Step 2 is to describe the input and output:

Input—array with five rows and seven columns
Output—number of negative values in the input array

Step 3 is to work a simple example by hand. Let the array be the following:

2	5	3	1	6	2	19
20	32	-4	2	7	13	2
8	25	-5	-4	18	4	9
28	-4	2	6	7	2	5
3	8	13	23	5	3	9

The number of negative values is 4.
Step 4 is to develop an algorithm, starting with the decomposition.

Decomposition

Count number of negative values in array k.
Return.

To count the number of negative values in a two-dimensional array, we need to use nested loops.

Pseudocode

```
negative_count_1(k):    negative_count_1 ← 0
                        i ← 1
                        while i ≤ 5
                             j ← 1
                             while j ≤ 7
                                   if k(i,j) < 0
                                           add 1 to negative_count_1
                                   add 1 to j
                             add 1 to i
                        return
```

Fortran Function

```
!------------------------------------------------------------------!
integer function negative_count_1(k)
!
!   This function counts the number of negative values
!   in an integer array with 5 rows and 7 columns.
!
!
!      Declare variables.
       implicit none
       integer :: i, j, k(5,7)
       intent (in) :: k
```

```
!
!     Count negative values.
      negative_count_1 = 0
      do i=1,5
         do j=1,7
            if (k(i,j) < 0) then
               negative_count_1 = negative_count_1 + 1
            end if
         end do
      end do
!
!     Function return.
      return
!
end function negative_count_1
!-------------------------------------------------------------------!
```

Step 5 is to test the program. A statement that might use the function in the main program, after filling an array `number` that has five rows and seven columns, is shown in the following group of statements:

```
      integer :: count, negative_count_1, number(5,7)
      ...
      count = negative_count_1(number)
```

. . . .

Problems can arise when you use variable dimensioning with arrays that have more than one dimension. The following is an example. First, we modify the `negative_count_1` function from Example 6-3 so that the number of rows and the number of columns are dummy arguments.

```
!-------------------------------------------------------------------!
integer function negative_count_2(k,n_rows,n_cols)
!
!  This function illustrates the problems that can
!  occur when using variable dimensioning with
!  two-dimensional arrays.
!
!
!     Declare variables.
      implicit none
      integer :: i, j, k(n_rows,n_cols), n_cols, n_rows
      intent (in) :: k, n_rows, n_cols
!
!     Count negative values.
      negative_count_2 = 0
      do i=1,n_rows
         do j=1,n_cols
            if (k(i,j) < 0) then
               negative_count_2 = negative_count_2 + 1
            end if
         end do
      end do
!
```

```
!     Function return.
      return
!
end function negative_count_2
!----------------------------------------------------------------!
```

If we use the following function reference, the correct value is stored in `count`:

```
integer :: count, n_cols=7, n_rows=5, &
           negative_count_2, number(5,7)
...
count = negative_count_2(number,n_rows,n_cols)
```

However, suppose we want to count the negative values using only the first two rows and the first two columns of the array `number`, as shown in the following diagram:

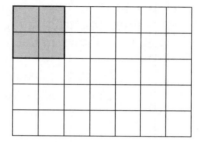

The following statements seem reasonable, but they are incorrect:

```
integer :: count, n_cols=2, n_rows=2, &
           negative_count_2, number(5,7)
...
count = negative_count_2(number,n_rows,n_cols)
```

These statements are incorrect because Fortran stores a two-dimensional array by columns. The array `number` is actually stored in memory, as shown in the following diagram:

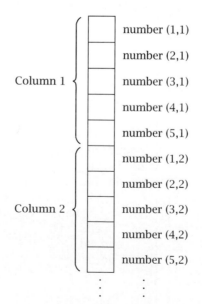

When we reference the array number in our function and specify that it has two rows and two columns, Fortran assumes that the values in number are stored in the following order:

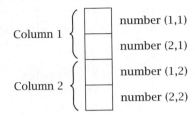

When we put the preceding two diagrams side-by-side, we can see that the array elements do not completely match:

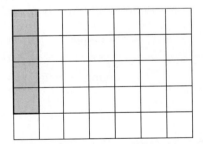

In fact, if we reference an array with two rows and two columns using the array number, the elements we are actually going to use are shaded in the following diagram:

This problem does not result in a compiler error, but the values used by the function are not those that we intend it to use, which can present a difficult error to find in a program. In general, you should not use variable dimensions with arrays that have more than one dimension. If you have an application that requires variable dimensioning with two-dimensional arrays, include arguments that specify an operational size and a dimensional size, where the operational size specifies the part of the array you are actually going to use, and the dimensional size specifies the size used in the array specification statement in the main program. We will use n_rows and n_cols for the size of the array we are going to use in the function and dim_rows and dim_cols for the dimensions used in the specification statement. These dummy arguments are used in the following function:

```
!------------------------------------------------------------!
integer function negative_count_3 &
               (k,n_rows,n_cols,dim_rows,dim_cols)
!
!  This function counts the number of negative values
!  in a subarray (of size n_rows by n_cols) of an
!  array (of size dim_rows by dim_cols).
!
!
!    Declare variables.
     implicit none
     integer :: dim_cols, dim_rows, i, j, &
                k(dim_rows,dim_cols), n_cols, n_rows
     intent (in) :: k, n_rows, n_cols, dim_rows, dim_cols
!
!    Count negative values.
     negative_count_3 = 0
     do i=1,n_rows
        do j=1,n_cols
           if (k(i,j) < 0)
              negative_count_3 = negative_count_3 + 1
           end if
        end do
     end do
!
!    Function return.
     return
!
end function negative_count_3
!------------------------------------------------------------!
```

The correct reference to count the negative values in the subarray of the array number using two rows and two columns is

```
     integer :: count, dim_cols=7, dim_rows=5, &
                n_cols=2, n_rows=2, number(5,7)
     ...
     count = negative_count_3(number,n_rows,n_cols, &
                               dim_rows,dim_cols)
```

Try It   Try this self-test to check your memory of some key points from Section 6-3. If you have any problems with the exercises, you should reread this section. The solutions are given at the end of this module.

For the following problems give the value of s assuming that a has the value 8.9, b has the value 3.1, c has the value 0.2, and d has the value − 5.5. The function test is the following:

```
!------------------------------------------------------------!
real function test(x,y,z)
!
!  This function returns either x+y, x-y, or x*z
!  depending on the values of x, y, z.
!
!
```

```
!     Declare variables.
      implicit none
      real :: x, y, z
      intent (in) :: x, y, z
!
      if (x > 10.0) then
         test = x + y
      else if (x > y) then
         test = x - y
      else
         test = x*z
      end if
!
!     Function return.
      return
!
end function test
!----------------------------------------------------------------!

1. s = test(a,b,c)
2. s = test(b,c,d)
3. s = test(abs(d),c,d)
4. s = test(3.0*c,c,d*d)
```

6-4 Application   OIL WELL PRODUCTION

Petroleum Engineering

The daily production of oil from a group of wells is entered into a data file each week for analysis. One of the reports that uses this data file computes the average production from each well and prints a summary of the overall production from this group of wells.

Write a Fortran program that will read the information from the data file and generate this report. Assume that the first line of the data file contains a date (month, day, and year of the first day of the week that corresponds to the production data). Each following line in the data file contains an integer identification number for the well and seven real numbers that represent the well's production for the week (in barrels). The number of wells to be analyzed varies from week to week, so a trailer line is included at the end of the file; this trailer contains the integer 99999, followed by seven zeros. You may assume that no well will have this integer as its identification number.

The report generated should have the following format:

```
Oil Well Production
Week of xx-xx-xx

Well Id      Average Production
             (in barrels)
xxxxx        xxx.xx
```

Include a final line at the end of the report that gives the total number of oil wells plus their overall average. Also, print the number of days that production is below an oil well's weekly average production.

Use the following set of data (stored in a file called `wells.dat`) for testing the program:

```
05  06  95
52      87      136     0       54      60      82      51
63      54      73      88      105     20      21      105
24      67      98      177     35      65      98      0
8       23      34      52      67      180     80      3
64      33      55      79      108     118     130     20
66      40      44      63      89      36      54      36
67      20      35      76      87      154     98      80
55      10      13      34      23      43      12      0
3       34      56      187     34      202     23      34
2       98      98      87      34      54      100     20
25      29      43      54      65      12      15      17
18      45      65      202     205     100     99      98
14      36      34      98      34      43      23      9
13      0       9       8       4       3       2       10
36      23      88      99      65      77      45      35
38      23      100     134     122     111     211     0
81      23      34      54      98      5       93      82
89      29      58      39      20      50      30      47
99      100     12      43      98      34      23      9
45      23      93      75      93      2       34      8
88      23      301     23      83      23      9       20
77      28      12      43      43      92      83      98
39      98      43      12      23      54      23      98
12      43      54      92      84      75      72      91
48      83      138     189     73      27      49      10
99999   0       0       0       0       0       0       0
```

1. Problem Statement

Generate a report on oil well production from daily production data. Give the average production for individual wells and the overall average.

2. Input/Output Description

Input—a data file with daily production data for a group of oil wells
Output—a report summarizing the oil production

3. Hand Example

For the hand example, we use the first five lines of the data file given at the beginning of this application, along with the trailer line:

```
05 06 95
52        87     136    0      54     60     82     51
63        54     73     88     105    20     21     105
24        67     98     177    35     65     98     0
8         23     34     52     67     180    80     3
99999     0      0      0      0      0      0      0
```

We use the date in the first line of the data file in our heading. Each individual data line in the report contains the well identification (the first value in each line) followed by the average. We compute the average from the seven daily values that follow the identification; we then count the number of days that the well had below-average production. We continue computing the individual averages until we reach the trailer signal 99999. At this point we average the individual wells and print the final summary lines:

```
Oil Well Production
Week of  5- 6-95

Well Id        Average Production
               (in barrels)
   52          67.14
   63          66.57
   24          77.14
    8          62.71

Overall average for  4 wells is 68.39
 15 days below weekly average production
```

4. Algorithm Development

Before we consider the steps in the algorithm, we first decide the best way to store the data. For instance, do we want to store it in a two-dimensional array, or do we want to store the individual oil production amounts in a one-dimensional array, or can we avoid arrays altogether?

We answer these questions by looking at the way in which we need to use the data. To compute the individual well average, we need a sum of the individual production amounts. Because all seven values are entered in the same data line, we must read them at the same time; thus, we need an array to store the seven daily production amounts for an individual well. Since we only need the data for one well at a time, a two-dimensional array is not necessary.

Decomposition

Generate report.
Print summary lines.

Initial Pseudocode

```
report:     read date
            read ID, oil production
            while ID ≠ 99999 do
                    compute individual average
                    print individual average
                    update overall average and well count
                    count the number of days with
                            below-average production
                    read next ID, oil production
            print summary information
```

After completing the decomposition and the initial refinement, we can determine if any of the operations should be written as functions. Computations that are repeated in an algorithm and steps that involve long computations are good candidates for functions. Program readability suffers when the details of some of the steps become long and tedious. Even though the steps may be performed only once, placing them in a function may make the program simpler.

For the oil well production problem, a function can be used to determine the number of days that production is less than the weekly average. This operation is needed only once in the main loop, but it involves several steps. The main program will be more readable if the steps are moved to a function. To be flexible, we write the function assuming that the data is in an array whose size is one of the function arguments.

Final Pseudocode

```
report:     Read date
            number of wells ← 0
            total oil ← 0.0
            below_average ← 0
            read ID, oil production
            while ID ≠ 99999 do
                    compute individual average
                    print individual average
                    add 1 to the number of wells
                    add individual well average to total oil
                    update the below_average count
                    read next ID, oil production
            print summary information

below_average(x, n, average):
            below_average ← 0
            k ← 1
            while k ≤ n
                    if x(k) < average
                            add 1 to below_average
                    add 1 to k
            return
```

Structure Chart

Fortran Program

```
!----------------------------------------------------------!
program oil_well_report
!
!  This program generates a report from the daily
!  production information for a set of oil wells.
!
!
!    Define and initialize variables.
     implicit none
     integer :: below_average, count=0, day, id, k, month, &
                n_days=7, n_wells=0, year
     real :: average, oil(7), total=0.0
!
!    Open data file and read date information.
     open(unit=12,file="wells.dat",status="old")
     read(12,*) month, day, year
!
!    Print report header.
     print*, "Oil Well Production"
     print 5, month, day, year
   5 format(1x,"Week of ",i2,"-",i2,"-",i2)
     print*
     print*, "Well Id      Average Production"
     print*, "             (in barrels)"
!
!    Read well data and print individual information.
     read(12,*) id, (oil(k),k=1,7)
     do while (id /= 99999)
        average = sum(oil)/7.0
        count = count + below_average(oil,n_days,average)
        print 10, id, average
  10    format(1x,i5,7x,f6.2)
        n_wells = n_wells + 1
        total = total + average
        read(12,*) id, (oil(k), k=1,7)
     end do
!
```

```
!     Print summary information.
      print*
      print 15, n_wells, total/real(n_wells)
   15 format(1x,"Overall average for ",i3," wells is ",f6.2)
      print, count
   20 format(1x,i3," days below weekly average production")
!
!     Close file.
      close(unit=12)
!
end program oil_well_report
!-----------------------------------------------------------------!
integer function below_average(x,n,average)
!
!  This function counts the number of values in the array x
!     (with n values) that are less than the value in average.
!
!
!     Declare variables.
      implicit none
      integer :: k, n
      real :: average, x(n)
      intent (in) :: x, n, average
!
!     Count values less than average.
      below_average = 0
      do k=1,n
         if (x(k) < average) below_average = below_average + 1
      end do
!
!     Function return.
      return
!
end function below_average
!-----------------------------------------------------------------!
```

5. Testing

Begin testing this program using a small data set such as the one in the hand-worked example. The output from this program using the data file given at the beginning of this application is given on the next page.

Could the last line of the data file contain only the trailer identification value 99999? (Are the seven zeros necessary in the last line of the data file?) If you try running the program without these last seven zeros, you will find that an execution error occurs because the program has run out of data. Because the identification number and the well production values are on the same line, we must read them with one read statement. Each time the read statement is executed, it reads eight values before control passes to the next statement. Thus, when the read statement reaches the last line in the data file, it needs seven values in addition to the 99999 value before it can test for the trailer value.

```
Oil Well Production
Week of  5- 6-95

Well Id        Average Production
               (in barrels)
    52            67.14
    63            66.57
    24            77.14
     8            62.71
    64            77.57
    66            51.71
    67            78.57
    55            19.29
     3            81.43
     2            70.14
    25            33.57
    18           116.29
    14            39.57
    13             5.14
    36            61.71
    38           100.14
    81            55.57
    89            39.00
    99            45.57
    45            46.86
    88            68.86
    77            57.00
    39            50.14
    12            73.00
    48            81.29

Overall average for  25 wells is  61.04
97 days below weekly average production
```

6-5 SUBROUTINE PROCEDURES

Subroutines are procedures written to perform operations that cannot be performed by a function. For example, if several values need to be returned from a procedure, a subroutine is used. A subroutine is also used for operations that do not compute values, such as reading the values in a data file.

Intrinsic Subroutines

Fortran 90 contains several intrinsic subroutines that are included within the compiler. The intrinsic subroutine that is used most frequently in engineering problem solutions is the one that allows us to generate pseudorandom numbers. The intrinsic subroutine random_number returns a random number between 0 and 1. The numbers are uniformly distributed across the interval between 0.0 and 1.0, which means that we are just as likely to get a value such as 0.4455 as we are to get a value such as 0.0090. This subroutine is called with the following reference:

```
call random_number(x)
```

If x is a real value, then a single random number between 0 and 1 is returned. If x is a real array, then each value in the array is filled with a different random number between 0 and 1. The following statement generates random numbers for the 10 values in a real array and then prints the random numbers:

```
real :: list(10)
...
call random_number(list)
print 5, list
5 format(1x,2f9.4)
```

Since different compilers are likely to determine the random numbers using different algorithms, the random numbers generated by these statements using one compiler will generally be different than random numbers generated by another compiler. The output of the statements using the Lahey Fortran 90 compiler is

```
0.4343   0.4544
0.9143   0.4826
0.7199   0.7314
0.8963   0.4702
0.0989   0.3267
```

When a program that contains the above statements is rerun, it gives the same set of random numbers. However, if the above statements appear in two different places in a program, the second set of statements generates a new set of random numbers. We will use this subroutine in the Application section that follows the discussion on user-written subroutines.

User-Written Subroutines

Many of the rules for writing and using subroutines are similar to those for functions. The following list of rules outlines the differences between subroutines and functions:

- A subroutine does not represent a value; thus, its name should be chosen for documentation purposes and not to specify a real or integer value.
- A subroutine is referenced with an executable statement whose general form is

```
call subroutine-name(argument list)
```

- The first line in a subroutine identifies it as a subroutine and includes the name of the subroutine and the argument list, as shown in this general form:

> subroutine *subroutine-name(argument list)*

- A subroutine uses the argument list not only for inputs to the subroutine but also for all values returned to the calling program. The subroutine arguments used in the `call` statement are the actual arguments, and the arguments used in the `subroutine` statement are the dummy arguments. The arguments in the `call` statement should match in type, number, and order those used in the subroutine definition.
- A subroutine may return one value, many values, or no value. Similarly, a subroutine may have one input value, many input values, or no input value. The declaration of all formal parameters should include an intent attribute, which can be `in`, `out`, or `inout`.
- Because the subroutine is a separate program, the arguments are the primary link between the main program and the subroutine. Thus, the choice of subroutine statement numbers and variable names is independent of those in the main program or in other procedures. The variables used in the subroutine that are not subroutine arguments are local variables, and their values are not accessible outside the subroutine.
- Be especially careful using multidimensional arrays in subroutines. It is generally advisable to pass both the dimensioned size and the operational size for arrays with two or more dimensions. (You may want to review the related discussion in Section 6.3.)
- A subroutine, like a function, should include a `return` statement to return control to the main program or to the procedure that called it. A subroutine also requires an `end subroutine` statement.
- In a flowchart the following special symbol is used to show that the operations indicated are performed in a subroutine:

- A subroutine may reference other functions or call other subroutines.

The following two examples develop subroutines.

EXAMPLE 6-4   ## Array Statistics

Information commonly needed from a set of data includes the average, the minimum value, and the maximum value. These values can be computed using intrinsic functions if we assume that the array is completely full of data. However, if the number of values of actual data in the array can vary, then we need to write a user-defined function to determine these statistics using only the valid information in the array. Write a subroutine that is called with this statement:

```
call statistics(x,n,x_average,x_minimum,x_maximum)
```

where n is the number of valid elements in the real array x.

SOLUTION

Step 1 is to state the problem clearly:

> Write a subroutine to determine the average,
> minimum, and maximum of an array of values.

Step 2 is to describe the input and output:

Input—an array of real values and the number of valid entries in the array
Output—the average, minimum, and maximum values from the array

Step 3 is to work a simple example by hand. These operations are all ones that we have used in examples before, however, so we do not repeat the hand examples here.

Step 4 is to develop an algorithm. Since we have already developed algorithms for determining averages, minimums, and maximums, the steps for the subroutine are straightforward.

Decomposition

Determine x_minimum, x_maximum, x_sum.
Compute x_average.

Pseudocode

```
statistics(x,n,x_average,x_minimum,x_maximum):
    x_sum ← x(1)
    x_minimum ← x(1)
    x_maximum ← x(1)
    k ← 2
    while k ≤ n
        if x(k) < x_minimum
            x_minimum = x(k)
        if x(k) > x_maximum
            x_maximum = x(k)
        add x(k) to x_sum
        add 1 to k
    x_average = x_sum/n
    return
```

Fortran Subroutine

```
!-----------------------------------------------------------------!
subroutine statistics(x,n,x_average,x_minimum,x_maximum)
!
!  This subroutine computes the average, minimum, and
!  maximum of a real array with n values.
!
!
```

```
!       Declare variables.
        implicit none
        integer :: k, n
        real :: x(n), x_average, x_maximum, x_minimum, x_sum
        intent (in) :: x, n
        intent (out) :: x_average, x_minimum, x_maximum
!
!       Set up loop to compute statistics.
        x_sum = x(1)
        x_minimum = x(1)
        x_maximum = x(1)
        do k=2,n
           if (x(k) < x_minimum) x_minimum = x(k)
           if (x(k) > x_maximum) x_maximum = x(k)
           x_sum = x_sum + x(k)
        end do
        x_average = x_sum/real(n)
!
!       Subroutine return.
        return
!
end subroutine statistics
!------------------------------------------------------------------!
```

Step 5 is to test the subroutine. We use a program that reads a set of exam scores and then calls the subroutine statistics to compute some statistics from the scores.

Fortran Program

```
!------------------------------------------------------------------!
program test_scores_1
!
!   This program reads a set of test scores and
!   then uses a subroutine to determine the average,
!   minimum, and maximum.
!
!
!     Declare variables.
      implicit none
      integer :: k, n
      real :: scores(100), score_ave, score_max, score_min
!
!     Prompt user to enter test scores.
      print*, "Enter number of test scores (<101)"
      read*, n
      print*, "Enter test scores:"
      read*, (scores(k), k=1,n)
!
!     Determine statistics.
      call statistics(scores,n,score_ave,score_min,score_max)
!
!     Print statistics.
      print*, "score average is ",score_ave
      print*, "minimum score is ",score_min
      print*, "maximum score is ",score_max
```

```
!
end program test_scores_1
!-----------------------------------------------------------------!
                    (subroutine statistics goes here)
!-----------------------------------------------------------------!
```

. . . .

EXAMPLE 6-5 ## Sort Subroutine

In Chapter 5 we developed the Fortran statements to sort a one-dimensional array. This operation is used so frequently that we will rewrite it in the form of a subroutine. To make it flexible, we use a variable in the argument list to specify the number of elements in the array. We suggest that you store this subroutine where it can be accessed easily.

SOLUTION

Step 1 is to state the problem clearly:

Write a subroutine to sort an array of data.

Step 2 is to describe the input and output:

Input—an array of real values and the number of valid entries in the array
Output—an array of the reordered data

Step 3 is to work a simple example by hand. Since we worked a hand example in Chapter 5 for the selection sort algorithm, refer to page 135 if you need to review it.
Step 4 is to develop an algorithm. Since we developed and coded this algorithm in Chapter 5, we need only modify the steps so that the solution is implemented as a subroutine.

Fortran Subroutine

```
!-----------------------------------------------------------------!
subroutine sort(x,n)
!
!  This subroutine sorts an array x into ascending
!  order.  x has n real values.
!
!
!    Declare variables.
     implicit none
     integer :: count, current, n, next_min
     real :: hold, x(n)
     intent (in) :: n
     intent (inout) :: x
!
!    Let current specify the current position
!    and count specify the number of values to sort.
     count = n
```

```
        do current=1,count-1
          next_min = current
!
!         Find next minimum.
          do k=current+1,count
             if (x(k) < x(next_min)) next_min = k
          end do
!
!         Switch the current position with next minimum.
          hold = x(current)
          x(current) = x(next_min)
          x(next_min) = hold
        end do
!
!     Subroutine return.
      return
!
end subroutine sort
!-----------------------------------------------------------------!
```

Step 5 is to test the subroutine. We will modify the previous program that read a set of test scores and printed some statistics from the scores so that the program also sorts and prints the test scores, five values per line. The structure chart is as follows:

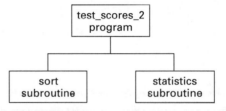

Fortran Program

```
!-----------------------------------------------------------------!
program test_scores_2
!
!   This program reads a set of test scores and
!   then uses a subroutine to determine the average,
!   minimum, and maximum.  It also prints the scores
!   in an ascending order.
!
!
!     Declare variables.
      implicit none
      integer :: k, n
      real :: scores(100), score_ave, score_min, score_max
!
!     Prompt user to enter test scores.
      print*, "Enter number of test scores (<101)"
      read*, n
      print*, "Enter test scores:"
      read*, (scores(k), k=1,n)
```

```
!
!     Determine statistics.
      call statistics(scores,n,score_ave,score_min,score_max)
!
!     Print statistics.
      print*, "score average is ", score_ave
      print*, "minimum score is ", score_min
      print*, "maximum score is ", score_max
!
!     Sort and print the test scores.
      call sort(scores,n)
      print 5, (scores(k), k=1,n)
    5 format(1x,5f9.2)
!
end program test_scores_2
```
!---!
                (subroutines statistics and sort go here)
!---!

   In the main program we specified that the list of values to be sorted had a
maximum of 100 values (because we dimensioned the array to 100 values).
The subroutine itself does not have any maximum on the array size; it can
sort any number of values as long as the values are part of an array that has
been properly defined in the main program or in the procedure that references the subroutine. Remember, however, that the original order of the
values is lost.

. . . .

Try It

Try this self-test to check your memory of some key points from Section
6-5. If you have any problems with these exercises, you should reread this
section. The solutions are given at the end of this module.

Questions 1 and 2 refer to the following program and subroutine. The
program generates an array of 10 numbers containing the integers 1
through 10. The subroutine modifies these numbers.

!---!
```
program quiz
!
!  This program tests your understanding of subroutines.
!
!
!     Declare variables.
      implicit none
      integer :: i, k(10)
!
!     Initialize array k.
      do i=1,10
         k(i) = i
      end do
!
!     Call subroutine.
      call modify(k,10)
```

```
!
!      Print array values.
       print*, "New values of k are:"
       print 5, (k(i), i=1,10)
     5 format(1x,5i5)
!
end program quiz
!-----------------------------------------------------------------!
subroutine modify(k,n)
!
!  This function modifies elements in an array.
!
!
!      Declare variables.
       implicit none
       integer :: i, k(n), n
       intent (in) :: n
       intent (inout) :: k
!
!      Set up loop to modify array element.
       do i=1,n
          k(i) = mod(k(i),4)
       end do
!
!      Subroutine return.
       return
!
end subroutine modify
!-----------------------------------------------------------------!
```

1. What is the program output?

2. What is the program output if the reference to the subroutine is replaced with the following statement?

```
                     call modify(k,8)
```

6-6 Application   SIMULATION DATA

Electrical Engineering

A routine to generate random numbers is useful in many engineering and
science applications. Most game programs use randomly generated numbers to make the program appear to have a mind of its own; it chooses
different actions each time the game is played. Programs that simulate
something, such as tosses of a coin or the number of people at a bank
window, also use random number generators. In this application we use
the intrinsic random number generator subroutine to simulate (or model)
noise, such as static, that might occur in a piece of instrumentation.

We want to develop a program that generates a data file containing a
sine wave plus noise using this equation:

$$f(t) = 2 \sin(2\pi t) + \text{noise}$$

for t = 0.0, 0.01, 0.02, and so on. Each value of the sine function is added to a random number produced by the random number generator. Since the sine wave (2 sin(2πt)) can vary from -2 to $+2$, and the random number generated can vary from 0 to 1, we expect this experimental signal to vary from -2 to 3. The signal should be stored in a data file called `signal.dat`, where each line of the data file contains the value of t and the corresponding signal value. The user should be asked to enter the number of points to generate, with a maximum of 500 points. The program should also print summary information giving the average value of the sine wave and its maximum and minimum values. To perform these computations, we will use the subroutine developed in the previous section.

 1. Problem Statement

Generate a data file that contains samples of the function f(t)=2 sin(2πt), with uniform random noise between 0.0 and 1.0 added to it. The data points are to be evaluated with t = 0.0, 0.01, and so on.

 2. Input/Output Description

Input—the number of function values to generate
Output—statistics on the function values and a data file `signal.dat`

 3. Hand Example

Using the random numbers given in Section 6.5, the first 10 data points of the file `signal.dat` are

$$f(0.00) = 2 \sin(2\pi \cdot 0.00) + 0.4343 = 0.4343$$
$$f(0.01) = 2 \sin(2\pi \cdot 0.01) + 0.4544 = 0.4670$$
$$f(0.02) = 2 \sin(2\pi \cdot 0.02) + 0.9143 = 1.1650$$
$$f(0.03) = 2 \sin(2\pi \cdot 0.03) + 0.4826 = 0.8574$$
$$f(0.04) = 2 \sin(2\pi \cdot 0.04) + 0.7199 = 1.2173$$
$$f(0.05) = 2 \sin(2\pi \cdot 0.05) + 0.7314 = 1.3494$$
$$f(0.06) = 2 \sin(2\pi \cdot 0.06) + 0.8963 = 1.6325$$
$$f(0.07) = 2 \sin(2\pi \cdot 0.07) + 0.4702 = 1.3218$$
$$f(0.08) = 2 \sin(2\pi \cdot 0.08) + 0.0989 = 1.0624$$
$$f(0.09) = 2 \sin(2\pi \cdot 0.09) + 0.3267 = 1.3984$$

 4. Algorithm Development

The only input is the number of signal values to generate. The program then generates the signal values and writes them in a data file. We also compute some statistics using the user-written subroutine developed in the previous section. Since the subroutine requires an array containing the signal values, we will store the signal values in an array, but we do not need to store the corresponding time values in an array.

Decomposition

Read number of values to generate.
Generate data values and write them to the file.
Determine and print statistics from data values.

Flowchart

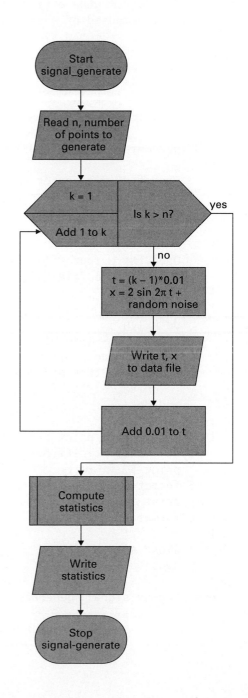

Structure Chart

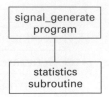

Fortran Program

```
!------------------------------------------------------------!
program signal_generate
!
!   This program generates a data file that contains a
!   signal composed of a sine wave plus random noise.
!
!
!     Declare variables.
      implicit none
      integer :: k, n
      real :: f(500), f_ave, f_max, f_min, noise, &
              pi, t
      parameter (pi=3.141593)
!
!     Prompt user to enter number of signal values.
      print*, "Enter number of signal values to generate:"
      print*, "(maximum of 500)"
      read*, n
!
!     Open the output file.
      open(unit=15,file="signal.dat",status="unknown")
!
!     Generate sine wave and add noise.
      do k=1,n
         t = (k - 1)*0.01
         call random_number(noise)
         f(k) = 2.0*sin(2.0*pi*t) + noise
         write(15,*)t, f(k)
      end do
!
!     Compute and print statistics
      call statistics(f,n,f_ave,f_min,f_max)
      print 5, f_ave
    5 format(1x,"Signal Average = ",f8.3)
      print 10, f_min, f_max
   10 format(1x,"Minimum = ",f8.3,", Maximum = ",f8.3)
!
!     Close space file.
      close(unit=15)
!
end program signal_generate
!------------------------------------------------------------!
                  (subroutine statistics goes here)
!------------------------------------------------------------!
```

5. Testing

To check this program, compare the first few data points in the data file created with those determined in the hand example. Remember that your values will be different than the ones we used if you are using a different compiler, because the random numbers generated will be different. One of the easiest ways to determine if this program is working properly is to plot the signal generated. The following plots show the sine wave without noise and then the sine wave with random noise added. These plots were generated using MATLAB.

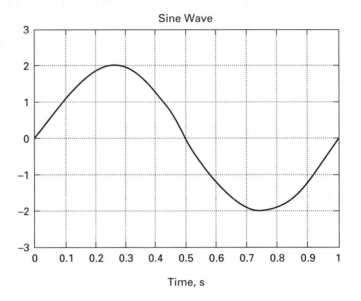

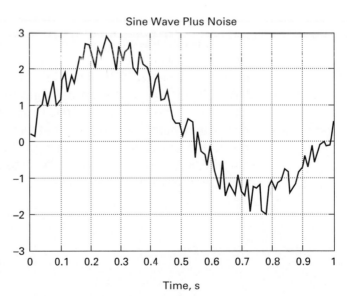

The average value printed by the program should be near 0.5, the minimum value should be near -2.0, and the maximum value should be near 3.0.

6-7   RECURSIVE PROCEDURES

Fortran 90 allows a procedure to refer to itself within the procedure. This special type of reference, called *recursion,* is a powerful technique that can be especially useful in solving certain types of problems. In recursion the current problem is defined in terms of a similar smaller problem, and the smaller problem is defined in terms of a still smaller similar problem, and so on until the solution of the smallest problem is finally reached.

The best way to understand recursion is with an example. A factorial computation occurs frequently in engineering and science applications. Recall that, by definition, n factorial, or n!, is computed as

$$n! = n \cdot (n - 1) \cdot (n - 2) \cdot \ldots \cdot 3 \cdot 2 \cdot 1$$

Thus

$$6! = 6 \cdot 5 \cdot 4 \cdot 3 \cdot 2 \cdot 1$$

However, note that we can also define 6! in terms of 5! as

$$6! = 6 \cdot 5!$$

Then we can define 5! in terms of 4! as in

$$5! = 5 \cdot 4!$$

Thus, a recursion process emerges as we define a factorial in terms of a smaller factorial, and that smaller factorial in terms of a still smaller factorial, until we get to 0!. By definition, 0! is equal to 1. This is a very critical step in recursion. We must eventually get to a stopping point, or recursion cannot be done with the computer. Once we get to the stopping point, we reverse the process by bringing that value back to the step that referenced it, compute a new value there, then go back again to the step that referenced that one, and so on until we are back to the original problem. The following diagram illustrates this process:

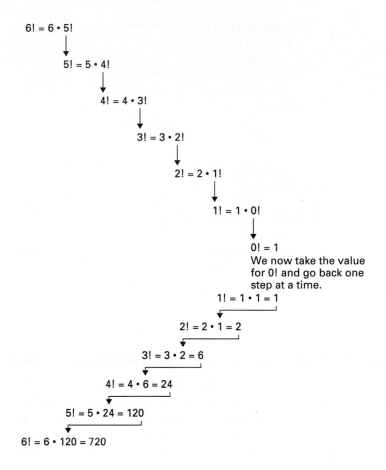

Note that the example recursively references the factorial function six times.

Fortran 90 allows recursive functions and recursive subroutines. These procedures must include the word `recursive` in the procedure heading. It is also necessary to include a `result` clause that allows the computer to distinguish between the use of the function name as a value and the function name as a reference to the function. A recursive function to compute a factorial is shown here:

```
!-----------------------------------------------------------------!
recursive function factorial(n) result (product)
!
!   This function recursively computes n!.  If n is
!   less than zero, a value of -1 is returned to
!   indicate an error has occurred.
!
!
!     Declare variables.
      implicit none
      integer :: n, product
      intent (in) :: n
!
```

```
!     Compute factorial recursively.
      if (n < 0) then
         product = -1
      else if (n == 0) then
         product = 1
      else
         product = n*factorial(n-1)
      end if
!
!     Function return.
      return
!
end function factorial
!------------------------------------------------------------------!
```

Recursive procedures are not commonly used in engineering problem solutions. In fact, an iterative solution is often preferred to a recursive solution because it takes less memory and less execution time. For some problems, however, a recursive solution is the simplest solution.

SUMMARY

This chapter presented both types of Fortran procedures—functions and subroutines. A function is a procedure that represents a single value. Fortran contains many intrinsic functions that compute values, such as trigonometric functions and logarithms. We can also write our own functions using separate procedures. A subroutine allows us to return multiple values, such as when we want to sort an array, or to return no values, such as when we want to print an error message.

Key Words

actual argument	procedure
dummy argument	recursion
external procedure	structure chart
function	subroutine
main program	variable dimensioning

Problems

We begin our problem set with modifications to programs developed earlier in this chapter. Give the decomposition, refined pseudocode or flowchart, and Fortran program for each problem.

Problem 1 modifies the oil well production report program developed in Section 6-4 on page 175.

1. Modify the oil production program so that it asks the user to enter the current date and then prints this date in the upper right corner of the first page of the report. Thus, the report will show the date that the report was run in addition to the date of the time period for which the data was collected.

Problem 2 modifies the signal generation program developed in Section 6-6 on page 188.

2. Modify the signal generation program so that it asks the user to enter the time step that is used between values of the signal f(t) that is generated.

For problems 3 and 4 write the function whose return value is described. The input to the function is an integer array k of 100 elements.

3. `positive_count(k)`, the number of values greater than or equal to zero in the array k.

4. `near_zero(k)`, the number of values of k between -0.0001 and 0.0001.

For problems 5-8 assume that you have a function subprogram `expression_1`, with input value x, to compute the following expression:

$$x^2 + \sqrt{1 + 2x + 3x^2}$$

Give statements that use this function to compute and print each of the following expressions, assuming that y has already been defined.

5. alpha $= \dfrac{6.9 + y}{y^2 + \sqrt{1 + 2y + 3y^2}}$

6. beta $= \dfrac{\sin y}{y^4 + \sqrt{1 + 2y^2 + 3y^4}}$

7. gamma $= \dfrac{2.3\,y + y^4}{y^2 + \sqrt{1 + 2y + 3y^2}}$

8. delta $= \dfrac{1}{\sin^2 y + \sqrt{1 + 2\sin y + 3\sin^2 y}}$

For problems 9-11 assume that `time`, a one-dimensional array of 100 real values, has already been filled with data.

9. Write a subroutine to determine the maximum value in the array `time` and to subtract that value from each value in the array.

10. Write a subroutine to determine the average of the values in the array `time`. Add 1.0 to all values above the average; subtract 1.0 from all values below the average.

11. Write a function to return a logical value of true if all the values in the array `time` are greater than zero; otherwise, the value returned should be false.

In problems 12-22 develop these programs and procedures. Use the five-step problem-solving process.

12. Assume that z, a two-dimensional array with five rows and four columns of real values, has already been filled with data. Write a sub-

routine that fills an array w (the same size as z) with a value based on the corresponding element in z, where

```
w(i,j) = 0.0 if z(i,j) = 0
w(i,j) = -1.0 if z(i,j) < 0
w(i,j) = 1.0 if z(i,j) > 0
```

13. Write a nonrecursive function to compute the factorial of an integer. Recall that the definition of a factorial is

$$0! = 1$$
$$n! = n \cdot (n-1) \cdot (n-2) \cdot \ldots \cdot 3 \cdot 2 \cdot 1$$

If n < 0, the function should return a value of − 1.

14. The cosine of an angle may be computed from the following series, where x is measured in radians:

$$\cos x = 1 - \frac{x^2}{2!} + \frac{x^4}{4!} - \frac{x^6}{6!} + \ldots$$

Write a function cosine whose input is an angle in radians. The function should compute and sum the first 10 terms of the series and return that approximation of the cosine. Use the factorial function developed in problem 13. (*Hint:* The alternating sign can be obtained by computing $(-1)^k$. When k is even, $(-1)^k$ is equal to $+1$; when k is odd, $(-1)^k$ is equal to -1.)

15. Write a main program that will produce a table with three columns. The first column (x) should contain angles from 0.0 to 3.1 radians in increments of 0.1 radians. The second column should contain the cosines of the angles as computed by the intrinsic function. The third column should contain the cosines as computed by the function in problem 14. Print the cosine values with an f10.7 format.

16. Engineering programs often utilize experimental data that have been collected and then stored in tabular form. The programs read the data and use it to generate plots, displays, or additional data. Examples of such data include wind tunnel force data on a new aircraft design, rocket motor thrust data, or automobile engine horsepower and torque data. The table of data that follows gives the thrust output of a new rocket motor as a function of time from ignition.

Time	Thrust
0.0	0.0
1.0	630.0
2.0	915.0
5.0	870.0
8.0	860.0
12.0	885.0
13.0	890.0
15.0	895.0

Time	Thrust
20.0	888.0
22.0	860.0
23.0	872.0
26.0	810.0
30.0	730.0
32.0	574.0
33.0	217.0
34.0	0.0

Write a function called thrust that computes the motor thrust for a rocket flight simulation program. The function should estimate the thrust for a specified time value using data from the preceding table. If the specified time value is not one of the times in the table, then the new data value can be determined from the tabulated values by using linear interpolation, which was discussed on page 128. For example, the estimated thrust at t = 2.2 seconds would be computed in the following manner:

$$\text{thrust} = 915.0 + \frac{2.2 - 2.0}{5.0 - 2.0} \cdot (870.0 - 915.0)$$

$$= 912.0$$

Assume that the function has two arguments: the two-dimensional real array with 16 rows and 2 columns, and the new value of time. If the time matches a time entry in the table, then the function should return the corresponding thrust value. If the time value does not match a time entry in the table, then the function should interpolate for a corresponding thrust value. Assume that the function will not be referenced if the time is less than zero or greater than 34.0.

17. Write a subroutine called bias that is called with the following statement:

```
call bias(x,n,y)
```

where x is an input array of 200 real values, n is an integer that specifies how many of the values represent actual data values, and y is an output array the same size as x whose values should be the values of x with the minimum value of the x array subtracted from each one. For example, if

$$x \quad \boxed{10 \mid 2 \mid 36 \mid 8}$$

then

$$y \quad \boxed{8 \mid 0 \mid 34 \mid 6}$$

Thus, the minimum value of y is always zero. (This operation is referred to as removing the bias in x or adjusting for bias in x.)

18. Write a subroutine that receives a two-dimensional real array x with 50 rows and 2 columns and returns the same array with the data reor-

dered. Sort the data such that the values in the second column are in ascending order. The values in the first column should correspond to the values in the second column. That is, the same values should be on a row together in both the original order and the new order, but the ordering of the rows may change.

19. Write a subroutine that receives a two-dimensional real array x with 50 rows and 2 columns and returns the same array with the data reordered. Sort the data such that the values in each column are in descending order.

20. Write a subroutine that receives a two-dimensional real array x with 50 rows and 4 columns and returns the same array with the data reordered. Sort the data such that the values in each row are in ascending order.

21. Write a subroutine that receives a two-dimensional real array x with 50 rows and 4 columns and returns the same array with the data reordered. An additional argument j is used to select a column that is to be sorted in descending order. The other values are not to be changed.

22. Write a subroutine that receives a two-dimensional real array x with 50 rows and 4 columns and returns the same array with the data reordered. An additional argument j is used to select a column that is to be sorted in descending order. The other values are to be changed so that values on the same row stay together in the new reordering.

7 Additional Data Types

| **Genetic Engineering** | Genetic engineering begins with a gene that produces a valuable substance, such as the human growth hormone. Enzymes are used to dissolve bonds to the neighboring genes, thus separating the valuable gene out of the DNA. This gene is then inserted into another organism, such as a bacterium, that will multiply itself along with the foreign gene. One step in discovering a valuable gene is identifying the sequence of amino acids in the protein that it produces. A protein sequencer is a sophisticated piece of equipment that can determine the order of amino acids making up a chainlike protein molecule, thus uncovering the identity of the gene that made it. Although there are only 20 different amino acids, protein molecules have hundreds of amino acids linked in a specific order.

INTRODUCTION

Standard Fortran 90 compilers must support five intrinsic, or built-in, data types: integer, real, logical, character, and complex. In addition, the compilers must support two types, or kinds, of real variables. real specifies one kind of real variable, and double precision specifies the other kind, which is assumed to store more significant digits than real variables store. This chapter presents the three data types that we have not yet discussed: character, double precision, and complex. Although we do not routinely use the data types discussed in this chapter, they are special features of Fortran that help make it a powerful language for engineering and scientific applications. We also present derived data types, which are structures composed of intrinsic data types.

7-1   CHARACTER DATA

The character data type allows us to read and analyze character data such as chemical formulas. Just like numbers, characters must be converted to *binary strings* to be stored in a computer's memory. Binary strings are composed of 1s and 0s. Several codes convert character information to binary strings, but most computers use *ASCII* (American Standard Code for Information Interchange) or *EBCDIC* (Extended Binary Coded Decimal Interchange Code). These binary codes represent each character by a binary string. Table 7-1 contains a few characters and their ASCII and EBCDIC equivalents. In addition, there are also standard codes for larger character sets, such as Unicode, which is an ISO (International Standards Organization) code.

You do not need to use binary codes to use the characters in your Fortran programs. However, you must be aware that the computer stores characters differently than the numbers used in arithmetic computations; that is, the integer number 5 and the character 5 are not stored in the same way. Thus, it is not possible to use arithmetic operations with character data even if the characters represent numbers.

We often refer to character data as *character strings,* because we usually refer to groups or lists of characters that go together. For example, we usually give one variable name to a chemical formula instead of giving a variable name to each character in the formula; thus, we might use the name water for the string 'H2O'. We can have character string constants that always represent the same information. Character string variables have names and represent character strings that may remain constant or may change. Generally, these character string constants and variables contain

Table 7-1   Binary Character Codes

Character	ASCII	EBCDIC
A	1000001	11000001
H	1001000	11001000
Y	1011001	11101000
3	0110011	11110011
+	0101011	01001110
$	0100100	01011011
=	0111101	01111110

Table 7-2   Fortran 90 Special Characters

Character	Character Name
	Blank
=	Equal
+	Plus
–	Minus
*	Asterisk
/	Slash
(	Left parenthesis
)	Right parenthesis
,	Comma
.	Decimal point or period
'	Apostrophe
:	Colon
!	Exclamation point
"	Quotation mark
%	Percent
&	Ampersand
;	Semicolon
<	Less than
>	Greater than
?	Question mark
$	Currency symbol

characters from the *Fortran character set,* which is composed of the 26 alphabetic letters, the 10 numeric digits, an underscore character (_), and the set of 21 special characters shown in Table 7-2. If other symbols are used in variable names, a program may not execute the same way on one computer as it does on another.

Character constants are always enclosed in either single or double quotation marks. These quotation marks are not counted when the length or number of characters in a constant is determined. If two consecutive single quotation marks (not a double quotation mark) are encountered within a character constant, they represent a single quotation mark. For example, the character constant for the word won't is "won''t". Two double quotes inside a constant also represent a single double quote character. The following list gives several examples of character constants and their corresponding lengths:

Constant	Length
"sensor 23"	9 characters
"Time and Distance"	17 characters
" $ amount"	9 characters
" "	2 characters
"08:40-13:25"	11 characters
""""""	2 characters

In addition to the standard character set and the set of special characters, Fortran 90 allows compilers to include additional character sets. These character sets can be used for special purposes and include symbols for mathematics, chemistry, and music. To incorporate additional languages, character sets such as Greek (commonly used in engineering and mathematical equations), Cyrillic (used in Russian and other Slavic languages), Hindi, Magyar (used in Hungarian), and Nihongo (used in Japanese) can be included.

A character string variable is defined with a specification statement whose general form is

```
character*n :: variable list
```

where n represents the number of characters in each character string. For instance, the statement

```
character*8 :: code, last_name
```

identifies `code` and `last_name` as variables containing eight characters each. Unlike numeric variable names, there is no significance to the first letter of a character variable's name. A variation of the `character` statement allows you to specify character strings of different lengths in the same statement, as shown:

```
character :: title*10, state*2
```

An array `name` that contains 50 character strings can be defined using either of the following statements:

```
character*4 :: name(50)

character :: name(50)*4
```

The preceding specifications reserve memory for 50 elements in the array `name`, where each element contains four characters. A reference to `name(18)` references a character string with four characters that is the 18th element of the array.

Character strings can also be used as arguments in procedures. The character string must be specified in a `character` statement in both the main program (or referencing procedure) and the procedure itself. A procedure can specify a character string as a formal argument without giving a specific length, as in

```
character*(*) :: string
```

This technique is similar to specifying the length of an array with an integer variable, as in

```
integer :: ssn(n)
```

It is also possible to define in a procedure an array of n variables, each of which contains a character string. To make the character string more flexible, its length does not have to be specified in the `character` statement. We use the following statement in the procedure to accomplish this flexibility:

```
character*(*) :: name(n)
```

Examples later in this section will illustrate how to use this statement.

Character I/O

When a character string is used in a list-directed output statement, the entire character string is printed. Blanks are usually inserted around the character string to separate it from other output on the same line, but this is system dependent. When a character string variable is used in a list-directed input statement, the corresponding data value must be enclosed in apostrophes or quotation marks. If the character string within the apostrophes or quotation marks is longer than the defined length of the character string variable, any extra characters on the right are ignored; if the character string is shorter than the length of the character string variable, the extra positions to the right are automatically filled with blanks. To print a character string in a formatted output statement, use a as the corresponding format specification to print the entire string.

EXAMPLE 7-1

Character I/O

Write a complete Fortran program to read an item description entered from the keyboard and then print the description. Assume that the length of the description is no more than 20 characters.

SOLUTION

We start with the decomposition.

Decomposition

Read description.
Print description.

The following refinement indicates the conversation with the user:

Pseudocode

```
Output:   print message to user to enter description
          read description
          print description
```

Translating these steps into Fortran results in the following program:

Fortran Program

```
!-----------------------------------------------------------------!
program description
!
!  This program reads and prints the
!  description of a piece of equipment.
!
!
```

```
!     Declare and initialize variables.
      implicit none
      character*20 :: item
!
!     Prompt the user to enter the description.
      print*, "Enter item description in quotes:"
      read*, item
!
!     Print the description.
      print 5, item
   5 format(1x,"item description is ",a)
!
end program description
!---------------------------------------------------------------!
```

A sample interaction with this program is

```
Enter item description in quotes:
"computer modem"
item description is computer modem
```

Note that the data entered was not 20 characters long. In this example the padding of blanks on the end is not noticeable in the output; however, if the output format had been

```
   5 format(1x,a," is the item description")
```

the interaction would have the following appearance:

```
Enter item description in quotes:
"computer modem"
computer modem         is the item description
```

Another interaction that could come from the original program is

```
Enter item description in quotes:
"digital oscilloscope with memory"
item description is digital oscilloscope
```

In this case the name exceeded the maximum number of characters specified for the description, so part of the data was lost. To avoid this situation, carefully choose the length of your character variables based on the maximum length expected. You can also tell the program user what length you are expecting. In this example you could use the following pair of print statements:

```
      print*, "Enter item description in quotes:"
      print*, "(maximum of 20 characters)"
```

. . . .

Character Operations

Although we cannot use character strings in arithmetic computations, we can assign values to character strings, compare two character strings, extract a substring of a character string, and combine two character strings into one longer character string.

Assign Values   Values can be assigned to character variables with the assignment statement and a character constant. The following statements initialize a character string array rank with the five abbreviations for freshman, sophomore, junior, senior, and graduate:

```
character*2 :: rank(5)
...
rank(1) = "FR"
rank(2) = "SO"
rank(3) = "JR"
rank(4) = "SR"
rank(5) = "GR"
```

If a character constant in an assignment statement is shorter in length than the character variable, blanks are added to the right of the constant. If the following statement were executed, rank(1) would contain the letter F followed by a blank:

```
rank(1) = "F"
```

If a character constant in an assignment statement is longer than the character variable, the excess characters on the right are ignored. Thus, the following statement,

```
rank(3) = "JUNIOR"
```

would store the letters JU in the character array element rank(3). These examples emphasize the importance of using character strings that are the same length as the variables used to store them.

One character string variable can also be used to initialize another character string variable, as shown:

```
character*4 :: grade_1, grade_2
...
grade_1 = "good"
grade_2 = grade_1
```

Both variables, grade_1 and grade_2, contain the character string "good".

Character strings can be initialized on the type statement or with data statements. The preceding examples can be performed as shown:

```
character :: grade_1*4 = "good", &
             grade_2*4 = "good", rank(5)*2
data rank /"FR", "SO", "JR", "SR", "GR"/
```

Table 7-3 Partial Collating Sequences for Characters

```
ASCII
    !"#$%&'()*+,-./0123456789:;<=>?@
    ABCDEFGHIJKLMNOPQRSTUVWXYZ
    [|]^_'abcdefghijklmnopqrstuvwxyz{|}~
EBCDIC
    .<(+|&!$*):_<%->?
    abcdefghijklmnopqr~stuvwxyz
    {ABCDEFGHI}JKLMNOPQR\STUVWXYZ
    0123456789
```

Compare Values An `if` statement can be used to compare character strings. Assuming that the variable `department` and the array `char` are character strings, the following are valid statements:

```
if (department == "EECE") count = count + 1

if (char(i) > char(i+1)) then
    call switch(i,char)
    call print(char)
end if
```

To evaluate a logical expression using character strings, first look at the length of the two strings. If one string is shorter than the other, the evaluation of the two strings will be made as if blanks were added to the right of the shorter string so that the strings are of equal length.

The comparison of two character strings of the same length is made from left to right, one character at a time. Two strings must have exactly the same characters in the same order to be equal.

A *collating sequence* lists characters from the lowest to the highest value. Partial collating sequences for ASCII and EBCDIC are given in Table 7-3. Although the ordering is not exactly the same, some similarities include

- Alphabetic letters are in order from a to z, or from A to Z. In ASCII, lowercase letters follow uppercase letters.
- Digits are in order from 0 to 9.
- Alphabetic letters and digits do not overlap; either digits precede letters or letters precede digits.
- The blank character is less than any letter or number.

The following is a list of several pairs of character strings, along with their correct relationships:

```
"A1"    < "A2"
"John"  < "Johnston"
"176"   < "177"
"three" < "two"
"$"     < "dollar"
```

If character strings contain only letters, their order from low to high is alphabetical, which is also called *lexicographic order*.

Extract Substrings   A *substring* of a character string is any string that represents a subset of the original string and maintains the original order. The following list contains all substrings of the string "Fortran":

```
"F"   "Fo"   "For"   "Fort"   "Fortr"   "Fortra"   "Fortran"
"o"   "or"   "ort"   "ortr"   "ortra"   "ortran"
"r"   "rt"   "rtr"   "rtra"   "rtran"
"t"   "tr"   "tra"   "tran"
"r"   "ra"   "ran"
"a"   "an"
"n"
```

Substrings are referenced by using the name of the character string, followed by two integer expressions in parentheses, separated by a colon. The first expression in parentheses gives the position in the original string of the beginning of the substring; the second expression gives the position of the end of the substring. If the string "Fortran" is stored in a variable language, some of its substring references are as shown:

Reference	Substring
language(1:1)	"F"
language(1:7)	"Fortran"
language(2:3)	"or"
language(7:7)	"n"

If the first expression in parentheses is omitted, the substring begins at the beginning of the string. If the second expression in parentheses is omitted, the substring ends at the end of the string; thus, language(5:) refers to the substring "ran".

The substring operation cannot operate correctly if the beginning and ending positions are not integers, are negative, or contain values greater than the number of characters in the substring. The ending position must also be greater than or equal to the beginning position of the substring.

The substring operator is a powerful tool, as the next example illustrates.

EXAMPLE 7-2   ## Character Count

A string of 50 characters contains encoded information. The number of occurrences of the letter s represents a special piece of information. Write a loop that counts the number of occurrences of the letter s, including both lowercase and uppercase occurrences.

SOLUTION

The loop index is used with the substring operator to allow us to test each character in the string:

```
character*50 :: code
integer :: count=0, k
...
do k=1,50
   if ((code(k:k) == "s").or.(code(k:k) == "S")
      count = count + 1
   end if
end do
```

. . . .

A reference to a substring can be used anywhere that a string can be used. For instance, if language contains the character string "Fortran", the following statement changes the value of language to "Formats":

$$language(4:7) = "mats"$$

If language contains "Formats", the following statement changes the value of language to "Formatt":

$$language(7:7) = language(6:6)$$

When modifying a substring of a character string with a substring of the same character string, the substrings must not overlap — that is, do not use language(2:4) to replace language(3:5). Also, recall that if a substring is being moved into a smaller string, only as many characters as are needed to replace the smaller string are moved from left to right; if the substring is being moved into a larger string, the extra positions on the right are filled with blanks.

Combine Strings   *Concatenation* is the operation of combining two or more character strings into one character string. It is indicated by two slashes (//) between the character strings to be combined. The following expression concatenates the constants "work" and "ed" into one string constant "worked":

$$"work"//"ed"$$

The next statement combines the contents of three character string variables mo, da, and yr into one character string and then moves the combined string into a variable called date:

$$date = mo//da//yr$$

If mo = "05", da = "15", and yr = "96", then date = "051596". Because concatenation represents an operation, it cannot appear on the left side of an equal sign.

Character Intrinsic Functions

A number of intrinsic functions are designed for use with character strings:

index locates specific substrings within a given character string.

len determines the length of a string and is used primarily in subroutines and functions that have character string arguments.

len_trim determines the length of a string without counting trailing blank characters.

char and ichar determine the position of a character in the collating sequence of the computer.

lge, lgt, lle, and llt allow comparisons to be made based on the ASCII collating sequence, regardless of the collating sequence of the computer.

index The index function has two arguments, both of which are character strings. The function returns an integer value that gives the position in the first string of the second string. Thus, if string_a contains the phrase "to be or not to be", index(string_a,"be") returns the value 4, which points to the first occurrence of the string "be". To find the second occurrence of the string, we can use the following statements:

```
integer :: k, j
character*18 :: string_a
...
k = index(string_a,"be")
j = index(string(k+1:),"be") + k
```

After executing these statements, k would contain the value 4, and j would contain the value 17. (Note that we had to add k to the second reference of index because the second use referred to the substring "e or not to be", and thus the second index reference returns a value of 13, not 17.) The value of index(string_a,"and") would be 0 because the string "and" does not occur in the first string string_a.

len The input to the function len is a character string; the output is an integer that contains the length of the character string. This function is useful in a procedure that accepts character strings of any length but needs the actual length. The statement in the procedure that allows character strings to be defined without specifying their length is

```
character*(*) :: a, b, string_a
```

This form can be used only in procedures. Example 7-3 uses both the len function and a variable string length in a procedure.

len_trim The input to the function len_trim is a character string; the output is an integer that contains the length of the character string without counting trailing blanks. For example, the value of len_trim("Hello! ") is 6, and the value of len("Hello! ") is 7.

EXAMPLE 7-3 ## Frequency of Blanks

Write a function that accepts a character string and returns a count of the number of blanks in the string.

SOLUTION

To make this function flexible, we write it so that it can be used with any size character string.

```
!------------------------------------------------------------------!
integer function blanks(x)
!
!   This function counts the number of blanks
!   in a character string x.
!
!
!     Declare variables.
      implicit none
      integer k
      character*(*) :: x
      intent (in) :: x
!
!     Count number of blanks.
      blanks = 0
      do k=1,len(x)
         if (x(k:k) == " ") blanks = blanks + 1
      end do
!
!     Function return.
      return
!
end function blanks
!------------------------------------------------------------------!
```

. . . .

EXAMPLE 7-4 ## Name Editing

Write a subroutine that receives three character strings, first, middle, and last, each containing 15 characters. The output of the subroutine is a character string 35 characters long that contains the first name followed by a blank, the middle initial followed by a period and a blank, and the last name. Assume that first, middle, and last have no leading blanks and no embedded blanks. Thus, if the variables are

first   Joseph
middle   Charles
last   Lawton

then the edited name would be

                    Joseph C. Lawton

SOLUTION

The solution to this problem is simplified by the use of the substring operation that allows us to look at individual characters and the index function that can be used to find the end of the first name. To the character string name we move the characters in first, then a blank, the middle initial, a period, another blank, and the last name. As you go through the solution, observe the use of the concatenation operation. Also, note that the move of the first name fills the rest of the character string name with blanks because first is shorter than the field to which it is moved.

```
!------------------------------------------------------------!
subroutine edit(first,middle,last,name)
!
!  This function edits a name to the form
!  first, middle initial, last.
!
!
!    Declare variables.
     implicit none
     integer k
     character*15 :: first, middle, last
     character*35 :: name
     intent (in) :: first, middle, last
     intent (out) :: name
!
!    Move first name.
     name = first
!
!    Move middle initial.
     k = index(name," ")
     name(k+1:k+3) = middle(1:1)//". "
!
!    Move last name.
     name(k+4:) = last
!
!    Subroutine return.
     return
!
end subroutine edit
!------------------------------------------------------------!
```

. . . .

char, ichar These functions refer to the collating sequence used in the computer. If a computer has 50 characters in its collating sequence, these characters are numbered from 0 to 49. For example, assume that the letter A corresponds to position number 12. The function char uses an integer argument that specifies the position of a desired character in the collating sequence, and the function returns the character in the specified position. The following statements print the character A:

```
n = 12
print*, char(n)
```

The ichar function is the inverse of the char function. The argument to the ichar function is a character variable that contains one character. The function returns an integer that gives the position of the character in the collating sequence. Thus, the output of the following statements is the number 12:

```
character*1 :: info="A"
...
print*, ichar(info)
```

Because different computers may have different collating sequences, the char and ichar functions can be used to determine the position of certain characters in the collating sequence.

lge, lgt, lle, llt This set of functions allows you to compare character strings based on the ASCII collating sequence. These functions become useful if a program is going to be used on a number of different computers and is using character comparisons or character sorts. The functions represent a logical value, true or false. Each function has two character string arguments, string_1 and string_2. The function reference lge(string_1,string_2) is true if string_1 is lexically greater than or equal to string_2; thus, if string_1 comes after string_2 in an alphabetical sort, this function reference is true. Remember, these functions are based on an ASCII collating sequence regardless of the sequence being used on the computer. The functions lgt, lle, and llt perform the comparisons of lexically greater than, lexically less than or equal to, and lexically less than, respectively.

Try It

Try this self-test to check your memory of some key points from Section 7-1. If you have any problems with the exercises, you should reread this section. The solutions are given at the end of this module.

For problems 1–10 give the substring referred to in each reference. Assume that a character string of length 35 called title has been initialized with the statements

```
character*35 :: title
...
title = "ten top engineering achievements"
```

1. title(1:20)
2. title(1:8)
3. title(9:19)
4. title(21:21)
5. title(9:)
6. title(:8)
7. title(:)
8. title(1:4)//title(21:)
9. title(5:8)//title(1:3)
10. title(9:16)//title(32:32)//"' '"//title(21:)

In problems 11–16 word is a character string variable of length 6. What is stored in word after each of the following statements?

11. word = "laser"
12. word = "fiber optics"
13. word = "CAD"//"CAM"
14. word = """"""""
15. word = " 12.48"
16. word = "genetic engineering"

PROTEIN MOLECULAR WEIGHTS

Genetic Engineering

Genetic engineering begins with a gene that produces a valuable substance, such as the human growth hormone discussed in the opening of this chapter. Enzymes are used to dissolve bonds to the neighboring genes, thus separating the valuable gene out of the DNA. This gene is then inserted into another organism, such as a bacterium, that will multiply itself along with the foreign gene.

One step in discovering a valuable gene is identifying the sequence of amino acids in the protein that it produces. A protein sequencer is a sophisticated piece of equipment that can determine the order of amino acids making up a chainlike protein molecule, thus uncovering the iden-

Table 7-4   Amino Acids

Amino Acid	Reference	Molecular Weight
Glycine	Gly	75
Alanine	Ala	89
Valine	Val	117
Leucine	Leu	131
Isoleucine	Ile	131
Serine	Ser	105
Threonine	Thr	119
Tyrosine	Tyr	181
Phenylalanine	Phe	165
Tryptophan	Trp	203
Aspartic	Asp	132
Glutamic	Glu	146
Lysine	Lys	147
Arginine	Arg	175
Histidine	His	156
Cysteine	Cys	121
Methionine	Met	149
Asparagine	Asn	132
Glutamine	Gln	146
Proline	Pro	116

tity of the gene that made it. Although there are only 20 different amino acids, protein molecules have hundreds of amino acids linked in a specific order.

In this problem we assume that the sequence of amino acids in a protein molecule has been identified and that we want to compute the molecular weight of the protein molecule. Table 7-4 lists the amino acids, their three-letter references, and their molecular weights. A data file named amino.dat contains 20 lines of data; each line of the data file contains the three-letter reference (enclosed in quotation marks) and the corresponding molecular weight.

Assume that another data file protein.dat contains the protein molecule characterizations in amino acids. The first line in the file contains the number of protein molecules in the file, and each following line contains a character string in quotation marks that contains the amino acid sequence. The maximum number of amino acids in a character string is 50. The program should determine and print the corresponding molecular weight for each protein. Also, we want to print an error message if an incorrect protein string is detected.

1. Problem Statement

Write a program that determines the molecular weight of a group of protein molecules containing only amino acids.

2. Input/Output Description

Input — a data file containing the information on amino acids and a data file containing the protein molecules
Output — the molecular weights of the protein molecules

3. Hand Example

Suppose that the protein molecule is the following:

LysGluMetAspSerGlu

The corresponding molecular weights for the amino acids are

147,146,149,132,105,146

Therefore, the protein molecular weight is 825.

4. Algorithm Development

We start the algorithm development with the decomposition.

Decomposition

Read and store amino acid data in arrays.
Read protein molecules and compute and print the molecular weight.

Pseudocode

```
molecular_weight:
            read amino strings and weights into arrays
            read the number of proteins, n
            k ← 1
            while k ≤ n
                    sum ← 0
                    read protein string
                    for each amino acid in the protein string
                            add corresponding weight to the sum
                    print protein string and sum
                    add 1 to k
```

The step to add the weight for each amino acid to the molecular weight sum requires comparing each amino acid string to the reference strings and then selecting the corresponding weight. We will implement this step in a subroutine in order to keep the main program easy to read. Since there are only a small number of amino acids to test, we use a do loop. If the amino acid is found, the corresponding weight will be moved to a variable named amino_wt; if the amino acid is not found, a value of zero will be returned in amino_wt.

```
amino_wt (reference,weights,string):
        amino_wt ← 0
        k ← 1
        while k ≤ 20
                if reference(k) = string
                        amino_wt = weights(k)
                add 1 to k
        return
```

There are several things to note about the following program. For example, we used the index function to determine the first blank in the protein string. This allowed us to determine the number of amino acids and also to print only the nonblank characters; otherwise, protein would require 150 output characters. Also, note that if we find an error in an amino acid, we continue evaluating the protein string. This allows us to catch multiple errors.

Fortran Program

```
!-------------------------------------------------------------!
program molecular_weights
!
!   This program reads character strings containing
!   amino acids from large protein molecules and
!   computes the corresponding molecular weights.
!
!
!     Declare variables.
      implicit none
      integer :: amino, amino_wt, blank, j, k, &
                 n_amino, n_char, n_proteins, &
                 start, sum, weights(20)
      character*3 :: reference(20)
      character*150 :: protein
      logical :: error
!
!     Open amino acid information file.
!     Read and store the information in arrays.
      open(unit=9,file="amino.dat",status="old")
      do k=1,20
         read(9,*) reference(k), weights(k)
      end do
!
!     Open protein molecules file and read the
!     number of molecules in the file.
      open(unit=10,file="protein.dat",status="old")
      read(10,*) n_proteins
!
!     Read protein molecule equation and compute
!     and print the corresponding weight.
      do j=1,n_proteins
!
         read(10,*) protein
         blank = index(protein," ")
         n_char = blank - 1
         n_amino = n_char/3
         sum = 0
```

```fortran
            if (mod(n_char,3) /= 0) then
              print 10, j
10            format(1x,"length error in protein ",i5)
              print*, protein(:blank)
              print*
            else
              error = .false.
              do k=1,n_amino
                 start = (k - 1)*3 + 1
                 amino = amino_wt(reference,weights, &
                                  protein(start:start+2))
                 if (amino /= 0) then
                    sum = sum + amino
                 else
                    print 12, k, j
12                  format(1x,"error in amino ",i3, &
                           ",  protein ",i3)
                    error = .true.
                 end if
              end do
!
              if (.not.error) then
                 print 15, protein(:blank), sum
15               format(1x,"protein: ",a/ &
                        1x,"molecular weight: ",i12)
              else
                 print*, protein(:blank)
              end if
            end if
         end do
!
!     Close files.
      close(unit=9)
      close(unit=10)
!
end program molecular_weights
!------------------------------------------------------------!
integer function amino_wt(reference,weights,string)
!
!     This function receives arrays containing the character
!     references and molecular weights for amino acids.
!     It returns the molecular weight of the amino acid.
!
!
!     Declare variables.
      implicit none
      integer :: k, weights(20)
      character*3 :: reference(20), string
      intent (in) :: reference, weights, string
!
!     Find matching amino acid string and
!     corresponding weight.
      amino_wt = 0
```

```
        do k=1,20
          if (string == reference(k)) amino_wt = weights(k)
        end do
!
!       Function return.
        return
!
end function amino_wt
!-------------------------------------------------------------!
```

5. Testing

The following protein.dat file was used for testing:

```
5
GlyIle
AspHisProGln
ThrTbrSerTrpLys
AlaValLeuValMet
LysGluMetAspSerGlu
```

The output for the test file was the following:

```
protein: GlyIle
molecular weight:            206
protein: AspHisProGln
molecular weight:            550
error in amino   2,   protein   3
ThrTbrSerTrpLys
protein: AlaValLeuValMet
molecular weight:            603
protein: LysGluMetAspSerGlu
molecular weight:            825
```

7-3 Double-Precision Data

With double-precision data, we can process numeric data more precisely than we could using previously discussed data types. Assume that a real variable can store seven significant digits; this means that the real variable will keep seven digits of accuracy, beginning with the first nonzero digit, in addition to remembering where the decimal point goes. A *double-precision value* can store more digits; the exact number of digits depends on the computer. For this discussion, assume that double-precision variables store 14 digits. The following table compares different precisions.

Value to Be Stored	Single Precision	Double Precision
37.6892718	37.68927	37.689271800000
−1.60003	−1.600030	−1.6000300000000
820000000487.	820000000000.	820000000487.00
18268296.300405079	18268290.	18268296.300405

Note that we are doubling the precision of our values, but we are not necessarily doubling the range of numbers that can be stored. The ranges of numbers for single-precision and double-precision values are also system dependent.

Many science and engineering applications use double-precision values to increase accuracy. For example, the study of solar systems, galaxies, and stars requires storing immense distances with as much precision as possible. Even economic models often need double precision. A model for predicting the gross national debt must handle numbers exceeding $1 trillion. With single precision we cannot store the values with an accuracy to the nearest dollar. With double precision we can use amounts up to $100 trillion and still have significant digits for all dollar amounts.

A double-precision constant is written in an exponential form, with a D in place of the E. Some examples of double-precision constants are

```
0.378926542D+04
1.4762D-02
0.25D+00
```

Always use the exponential form with the letter D for double-precision constants, even if the constant uses seven or fewer digits of accuracy; otherwise, you may lose some accuracy because a fractional value that can be expressed exactly in decimal notation may not be expressed exactly in binary notation.

The general form of the specification statement for double-precision variables is

```
double precision :: variable list
```

A double-precision array is specified as

```
double precision :: temperatures(50)
```

Double-Precision I/O

Double-precision variables can be used in list-directed output in the same manner that real values are listed. The only distinction is that more digits of accuracy can be stored in a double-precision value; therefore, more digits of accuracy can be written from a double-precision value.

In formatted input and output, double-precision values may be referenced with the f or e format specifications. Another specification, dw.d, may also be used; it functions essentially like the e specification, but the d emphasizes that it is being used with a double-precision value. In output the value in this exponential form is printed with a D instead of an E. Thus, if the following statements were executed,

```
      double precision :: dx
      ...
      dx = 1.66587514521D+00
      print 10, dx
   10 format(1x,"value is ",d17.10)
```

the output would be

```
value is   0.1665875145D+01
```

EXAMPLE 7-5

Solar Distances

Assume that a character array has been filled with the names of planets, moons, and other celestial bodies. A corresponding array has been filled with the distances of these objects from the sun in millions of miles. Both arrays have been defined to hold 200 values, and an integer n specifies how many elements are actually stored in the arrays. The array name is an array of character strings of length 20, and the array distance is a double-precision array. Give the statements to print these names and distances.

SOLUTION

Before printing the data in the arrays, we print a heading for the data. Then a loop is executed n times and is used to print the object name and its corresponding distance from the sun in millions of miles. We use a d format for the output because the set of distances may cover a large range of values.

```
integer :: k, n
character*20 :: name(200)
double precision :: distance(200)
...
print*, "Solar Objects and Distances from the Sun"
print*, "                    (millions of miles)"
print*
do k=1,n
   print 5, name(k), distance(k)
 5 format(1x,a,2x,d15.8)
end do
```

A sample output from these statements is

```
Solar Objects and Distances from the Sun
                    (millions of miles)

Jupiter                0.43863717D+03
Mars                   0.14151751D+03
Saturn                 0.88674065D+03
Venus                  0.67235696D+02
Pluto                  0.36662718D+04
Uranus                 0.17834237D+04
Mercury                0.35979176D+02
Neptune                0.27944448D+04
Earth                  0.92961739D+02
```

. . . .

Double-Precision Operations

When an arithmetic operation is performed with two double-precision values, the result is double precision. If an operation involves a double-precision value and a single-precision value or an integer, the result is double-precision. In such a mixed-mode operation, do not assume that the other value is converted to double precision; instead, think of the other value as being extended in length with zeros. To illustrate, the first two assignment statements that follow yield exactly the same values; the third assignment statement, however, adds a double-precision constant to dx and yields the most accurate result of the three statements:

```
double precision dx, dy1, dy2, dy3
...
dy1 = dx + 0.3872
dy2 = dx + 0.387200000000000
dy3 = dx + 0.3872D+00
```

The most accurate way to obtain a constant that cannot be written in a fixed number of decimal places is to perform a double-precision operation that yields the desired value. For instance, to obtain the double-precision constant $\frac{1}{3}$, use the following expression:

```
1.0D+00/3.0D+00
```

Using double-precision values increases the precision of our results, but we pay a price for this additional precision: The execution time for computations is longer, and more memory is required.

Double-Precision Intrinsic Functions

If a double-precision argument is used in a generic function, the function value is also double precision. Many of the common intrinsic functions for real numbers can be converted to double-precision functions by preceding the function name with the letter d. For instance, dsqrt, dabs, dmod, dsin, dexp, dlog, and dlog10 all require double-precision arguments and yield double-precision values. Since the generic function and double-precision function perform the same operation with double-precision arguments, we recommend using the generic function name for consistency. Double-precision functions can also be used to compute constants with double-precision accuracy. For instance, the following statements compute π with double-precision accuracy:

```
double precision pi
...
pi = 4.0D+00*atan(1.0D+00)
```

(Recall that $\pi/4$ is equal to the arctangent of 1.0.)

Two functions, dble and dprod, are specifically designed for use with double-precision variables. dble converts a real argument to a double-precision value by adding zeros. dprod has two real arguments and returns the double-precision product of the two arguments.

Try It

Try this self-test to check your memory of some key points from Section 7-3. If you have any problems with the exercises, you should reread this section. The solutions are given at the end of this module.

In problems 1–6 show how to represent these constants as double-precision constants.

1. 0.75   2. 1.3
3. 1/9   4. 5/6
5. −10.5   6. 3.0

In problems 7–9 show the output of the following statements if dx is equal to 0.00786924379, where dx is a double-precision variable.

7.   print 4, dx   8.   print 5, dx
    4 format(1x,d14.5)   5 format(1x,d12.3)
9.   print 6, dx
    6 format(1x,d12.7)

7-4 COMPLEX DATA

With complex data we can represent numbers that have a real portion and an imaginary portion. Complex numbers are needed to solve many problems in science and engineering, particularly in physics and electrical engineering. Therefore, Fortran includes a special data type for complex variables and constants. (Recall that *complex values* have the form a + bi, where i is $\sqrt{-1}$ and a and b are real numbers. Thus, the real part of the number is represented by a, and the imaginary part of the number is represented by b.) These complex values are stored as an ordered pair of real values that represents the real and the imaginary portions of the value.

A complex constant is specified by two real constants separated by a comma and enclosed in parentheses. The first constant represents the real part of the complex value; the second constant represents the imaginary part of the complex value. Thus, the complex constant 3.0 + 1.5i is written in Fortran as the complex constant (3.0, 1.5).

The specification statement for complex variables has the general form

```
complex :: variable list
```

A complex array is specified as

```
complex :: cx(100)
```

Complex I/O

A complex value in list-directed output is printed as two real values separated by a comma and enclosed in parentheses. Two real values are read for each complex value in a list-directed input statement.

Table 7-5   Complex Arithmetic Operations

Operation	Result
$c_1 + c_2$	$(a_1 + a_2) + i\,(b_1 + b_2)$
$c_1 - c_2$	$(a_1 - a_2) + i\,(b_1 - b_2)$
$c_1 \cdot c_2$	$(a_1 a_2 - b_1 b_2) + i\,(a_1 b_2 - a_2 b_1)$
$\dfrac{c_1}{c_2}$	$\dfrac{a_1 a_2 - b_1 b_2}{a_2{}^2 + b_2{}^2} + i\,\dfrac{a_2 b_1 - b_2 a_1}{a_2{}^2 + b_2{}^2}$
$\lvert c_1 \rvert$	$\sqrt{a_1{}^2 + b_1{}^2}$
e^{c_1}	$e^{a_1}\cos b_1 + i\,e^{a_1}\sin b_1$
$\cos c_1$	$1 - \dfrac{c_1{}^2}{2!} + \dfrac{c_1{}^4}{4!} - \dfrac{c_1{}^6}{6!} + \ldots$

(Note that $c_1 = a_1 + b_1 i$ and $c_2 = a_2 + b_2 i$.)

For formatted output a complex value is printed with two real specifications. The real part of the complex value is printed before the imaginary portion. It is good practice to enclose the two parts printed in parentheses and separate them by a comma or print them in the a + bi form. Both forms are illustrated in the following statements:

```
complex :: cx, cy
...
cx = (1.5, 4.0)
cy = (0.0, 2.4)
print 5, cx, cy
5 format(1x,"(",f4.1,",",f4.1,")"/ &
         1x,f4.1," + ",f4.1," i")
```

The output from the print statement is

```
( 1.5, 4.0)
0.0 + 2.4 i
```

Complex Operations

When an arithmetic operation is performed between two complex values, the result is also a complex value. In an expression containing a complex value and a real or integer value, the real or integer value is converted to a complex value whose imaginary part is zero. Expressions containing both complex values and double-precision values are not allowed.

The rules for complex arithmetic are not as familiar as those for integers or real values. Table 7-5 lists the results of a group of operations on two complex numbers c_1 and c_2, where c_1 has the value $a_1 + b_1 i$, and c_2 has the value $a_2 + b_2 i$.

Complex Intrinsic Functions

If a complex value is used in one of the generic functions, such as sqrt, abs, sin, cos, exp, or log, the function value is also complex. The functions csqrt, cabs, csin, ccos, cexp, and clog are all intrinsic functions with complex arguments. These function names begin with the letter c to emphasize that they are complex functions.

Four functions, real, aimag, conjg, and cmplx, are specifically designed for use with complex variables: real yields the real part of its complex argument; aimag yields the imaginary part of its complex argument; conjg converts a complex number to its conjugate, where the conjugate of (a + bi) is (a − bi); cmplx converts two real arguments, a and b, into a complex value (a + bi). Note that, while (2.0,1.0) is equal to the complex constant 2.0 + 1.0i, we must use the expression cmplx(a,b) to specify the complex variable a + bi; the expression (a,b) by itself does not represent a complex variable.

EXAMPLE 7-6

Quadratic Formula

The roots of a quadratic equation with real coefficients may be complex. Give the statements to compute and print the two roots of a quadratic equation, given the coefficients a, b, and c and assuming that a is not equal to zero:

$$ax^2 + bx + c = 0$$

$$x_1 = \frac{-b + \sqrt{b^2 - 4ac}}{2a} \qquad x_2 = \frac{-b - \sqrt{b^2 - 4ac}}{2a}$$

SOLUTION

```
      complex :: discriminant, root_1, root_2
      ...
      discriminant = cmplx(b*b - 4.0*a*c,0)
      root_1 = (-b + sqrt(discriminant))/(2.0*a)
      root_2 = (-b - sqrt(discriminant))/(2.0*a)
      print*, "Roots to the quadratic equation are:"
      print 5, root_1, root_2
    5 format(1x,f5.2," + ",f5.2," i",4x,f5.2," + ",f5.2," i")
```

Two sets of sample output are shown, one in which both roots are real and one in which both roots are complex:

```
      Roots to the quadratic equation are:
       1.56 +  0.00 i     2.00 +  0.00 i

      Roots to the quadratic equation are:
       2.36 +  2.45 i     2.36 + -2.45 i
```

Try It

Try this self-test to check your memory of some key points from Section 7-4. If you have any problems with the exercises, you should reread this section. The solutions are given at the end of this module.

In problems 1–6 compute the value stored in cx if cy is equal to 2i and cz is equal to 1+2i. Assume that cx, cy, and cz are complex variables.

1. cx = cy + 2*cz   2. cx = cz - cy
3. cx = conjg(cy)   4. cx = real(cz) + aimag(cy)
5. cx = cmplx(3.5,-1.5)   6. cx = abs(cy + cz)

In problems 7 and 8 show the output of the following print statements. Assume that cx has the value 2i.

7.   print 5, cx
   5 format(1x,f5.1," + ",f5.1," i")
8.   print 6, cx
   6 format(1x,"(",f6.2,",",f6.2,")")

7-5 DERIVED DATA TYPES

We have now discussed the intrinsic data types available in Fortran: integers, real values, double-precision values, logical values, characters, and complex numbers. We can define a simple variable or an array of variables that represent information using one of these data types. However, we cannot define an array that contains more than one data type. For example, we cannot define an array that contains real values in one column and characters in another column.

A new type of structure called a *derived data type* allows us to group information of different data types. For example, suppose that we want to define a data structure that will contain a character string that represents the symbol for a chemical element and a real number that represents its atomic weight. The following definition defines this new data type and gives it the name atom:

```
type atom
    character*3 :: symbol
    real :: weight
end type atom
```

This definition does not reserve any memory; it just defines the structure. To define variables using this derived data type, we would follow the type definition with variable declarations, as shown below:

```
! Define atom data type.
type atom
    character*3 :: symbol
    real :: weight
end type atom
!
! Declare variables.
type(atom) :: hydrogen, oxygen
```

To assign values to the components of a derived data type, use the structure identifier followed by a percent sign (%) followed by the component identifier. For example, the following statements give values to the components of the two derived data types defined above:

```
hydrogen%symbol = "H"
hydrogen%weight = 1.00794
oxygen%symbol = "O"
oxygen%weight = 15.9994
```

Then, to compute the molecular weight of water (H_2O), we could use the following statement:

```
water = 2*hydrogen%weight + oxygen*weight
```

An array of `atom` data types can be defined as shown in this statement:

```
type(atom) :: periodic_table(100)
```

This structure defines an array called `periodic_table`. Each element in the array contains a character string and a real value. The following statements assign values to the 17th element in the array:

```
periodic_table(17)%symbol = "Cl"
periodic_table(17)%weight = 35.4527
```

Derived data types are not used frequently in engineering problem solutions, but they can be very useful in representing some types of information. In Chapter 8 we will use derived data types to implement linked lists.

SUMMARY

With the data types presented in this chapter, we now have a number of choices for defining our variables and constants. For numeric data we can select integers, real values, or double-precision values. We have character data for information that is not going to be used in numeric computations. For special applications we have complex variables. In addition to being able to choose the proper data type for our data, we also have special intrinsic functions and operations for simplifying our work with the data.

Key Words

ASCII
binary string
character string
collating sequence
complex value
concatenation

derived data type
double-precision value
EBCDIC
Fortran character set
lexicographic order
substring

Problems

This problem set begins with modifications to programs developed earlier in this chapter. Give the decomposition, refined pseudocode or flowchart, and Fortran program for each problem.

Problems 1–5 modify the program `molecular_weights` given in Section 7-2 on page 214, which computes the molecular weight of a large protein molecule.

1. Modify the molecular weight program so that it includes a subroutine that is used to print the contents of the file `amino.dat` at the beginning of the program.
2. Modify the molecular weight program so that it prints a final line with the maximum and minimum protein molecular weights.
3. Modify the molecular weight program so that it prints a final line with the average protein molecular weight.
4. Modify the molecular weight program so that it prints the number of amino acids in each protein molecule.
5. Modify the molecular weight program so that it prints the maximum and minimum numbers of amino acids in the protein molecules.

In problems 6–12 develop programs and modules using the five-step problem-solving process.

6. Write a complete program that reads a data file `address.dat` containing 50 names and addresses. The first line for each person contains the first name (10 characters), the middle name (6 characters), and the last name (21 characters). The second line contains the address (25 characters), the city (10 characters), the state abbreviation (2 characters), and the zip code (5 characters). Each character string is enclosed in quotes in the file. Print the information in the following label form:

> First Initial. Middle Initial. Last Name
> Address
> City, State Zip

Skip four lines between labels. The city should not contain any blanks before the comma that follows it. A typical label might be

> J. D. Doe
> 117 Main St.
> Taos, NM 87166

For simplicity assume there are no embedded blanks in the individual data values. For example, San Jose would be entered as SanJose and printed in the same manner.

7. Write a complete program to read a double-precision value from the terminal. Compute the sine of the value using the following series:

$$\sin x = x - \frac{x^3}{3!} + \frac{x^5}{5!} - \frac{x^7}{7!}$$

Continue using terms until the absolute value of a term is less than 1.0D−09. Print the computed sine and the value obtained from the sine function for comparison.

8. A palindrome is a word or piece of text that is spelled the same forward and backward. The strings "radar" and "ABLE ELBA" are both palindromes. Write a logical function `palindrome` that receives a character variable x and an integer that specifies the length of the variable x. The function should be true if the character array is a palindrome; otherwise, it should be false.

9. Write a subroutine `alphabetize` that receives an array of 50 letters. The subroutine should alphabetize the list of letters.

10. Modify the subroutine in problem 9 to remove duplicate letters and to add blanks at the end of the array for the letters removed.

11. Write a subroutine that receives a piece of text called `prose` that contains 200 characters. The subroutine should print the text in lines of 30 characters each. Do not split words between two lines. Do not print any lines that are completely blank.

12. Write a complete program that reads and stores the following two-dimensional array of characters:

```
ATIDEB
LENGTH
ECPLOT
DDUEFS
OUTPUT
CGDAER
HIRXJI
KATIMN
BHPARG
```

Now read the 11 strings that follow and find the same strings in the preceding array. Print the positions of characters of these hidden words that may appear forward, backward, up, down, or diagonally. For instance, the word EDIT is located in positions (1,5), (1,4), (1,3), and (1,2).

Hidden Words
PLOT
STRING
CODE
TEXT
EDIT
READ
LENGTH
GRAPH
INPUT
BAR
OUTPUT

8 An Introduction to Pointers

Environmental Engineering Each year the migration of the whooping cranes is a special event for wildlife conservationists. Information on sightings is received by the Wildlife Service. This information, initially a few pieces of data, grows to hundreds of pieces of data as the cranes complete their migration. With advances in technology, the migration of animals can also be tracked using satellites that record information from transmitters attached either to collars placed on the animals or to wings. This technology is being used to study the migration habits of a wide variety of animals and birds. Satellite tracking is also being used to map the flight patterns of birds such as the albatross.

INTRODUCTION

A pointer variable, also called a *pointer,* is an alias to another variable. The use of pointer variables allows us to develop new techniques for solving problems, and also allows us to implement new types of data structures such as linked lists. In this chapter we introduce pointer variables and present a number of simple examples to illustrate the relationship between pointers and the variables that they *alias,* or to which they point. We then present additional examples using arrays of pointers and linked lists.

8-1 POINTER VARIABLES

Variables that will be used as pointers must be defined with a `pointer` attribute. Furthermore, when a pointer is defined, the type of variable to which it will point must also be defined. Thus, a pointer defined to point to an integer cannot also be used to point to a real variable. In addition, a variable that will be aliased, or pointed to, by a pointer must be given a `target` attribute. The following statements define two integer variables and a pointer to an integer variable:

```
real, target :: a, b
real, pointer :: ptr
```

These statements do not specify the initial values for a, b, and ptr. Thus, the memory snapshot specified by these statements is shown below, with an arrow used to indicate that ptr is a pointer variable:

Pointer Assignment

To assign a value to a pointer variable, we use a new operator specified by the characters =>. Thus, to specify that the pointer ptr in the previous example should alias the variable a, we use the following assignment statement:

```
ptr => a
```

The memory snapshot after this statement is executed is the following:

Note that the value aliased by a pointer does not have to be initialized.
Consider this set of statements:

```
integer, target :: a=2, b=8
integer, pointer :: ptr
...
ptr => a
```

The memory snapshot after this statement is executed is the following:

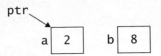

The pointer assignment statement in this example is read as "a is aliased by ptr" or "a is pointed to by ptr." If we execute the statement

ptr => b

the memory snapshot is now the following:

Since a pointer is an alias to another variable, using the pointer name is the same as using the name of the variable to which the pointer is aliased. To illustrate, assume that the current memory snapshot is the one shown above; then assume that the following statement is executed:

a = ptr

In executing this statement, the variable ptr is an alias to the variable b, and thus this statement assigns the value of b to the variable a. The resulting memory snapshot is the following:

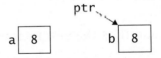

In the previous examples, the pointer ptr was either undefined or it was associated with a target variable. A pointer can also be a *null pointer,* which means that it has been initialized but does not point to a target variable; initializing a pointer to a null value is similar to initializing an integer to a value of zero. A pointer is given a null value, or is nullified, by the following statement:

nullify *(pointer)*

The memory snapshot for a null pointer often uses a special symbol to indicate a null value, as shown below:

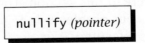

Consider these statements:

```
integer, target :: a=-2, b=5
integer, pointer :: ptr_1, ptr_2
...
ptr_1 => a
nullify(ptr_2)
```

After these statements are executed, the corresponding memory snapshot is the following:

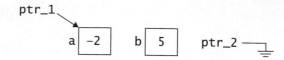

Now assume that the following statement is executed:

```
ptr_2 => b
```

The memory snapshot is

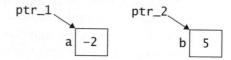

Assume the following statement is then executed:

```
ptr_1 = ptr_2
```

Since `ptr_1` is an alias for a, and since `ptr_2` is an alias for b, this statement is equivalent to

```
a = b
```

and thus the new memory snapshot is

Try It

Try this self-test to check your memory of some key points from this part of Section 8-1. If you have any problems with the exercises, you should reread this section. The solutions are given at the end of this module.

For problems 1–5 give memory snapshots after each set of statements is executed:

```
1. integer, target :: a=3, b=5, c
   integer, pointer :: ptr_1, ptr_2
   ...
   ptr_1 => b
   ptr_2 => a
   c = a + ptr_1
```

```
2. integer, target :: a=3, b=5, c=3
   integer, pointer :: ptr_1, ptr_2
   ...
   ptr_1 => b
   ptr_2 => c
   b = ptr_1*ptr_2 - c

3. integer, target :: a=3, b=5, c
   integer, pointer :: ptr_1, ptr_2
   ...
   ptr_1 => b
   ptr_2 => ptr_1
   if (ptr_1 > 3) c = ptr_1

4. integer, target :: a=3, b=5
   integer, pointer :: ptr_1, ptr_2
   ...
   ptr_1 => b
   a = ptr_1
   nullify(ptr_2)
   ptr_1 => ptr_2

5. integer, target :: a=3, b=5, c=8
   integer, pointer :: ptr_1, ptr_2
   ...
   ptr_1 => b
   ptr_2 => c
   ptr_1 = ptr_2 + a
```

Pointers can also be assigned using the `allocate` statement, which has the following general form:

> `allocate` *(pointer)*

When this statement is executed, a target memory location is assigned to the pointer. The only way to access the target memory location is through the pointer, because the location does not also have an identifier assigned to it. The following statements assign a pointer to a target memory location with the declaration statement and then assign a different pointer to a different target memory location using the `allocate` statement:

```
integer, target :: x
integer, pointer :: ptr_1, ptr_2
...
ptr_1 => x
allocate(ptr_2)
```

The memory snapshot after these statements are executed is the following:

Neither memory location has been assigned a value. One memory location can be accessed using either x or its alias ptr_1; the other memory location can be accessed only through ptr_2.

The allocate statement is used to assign pointer targets when implementing linked lists. This important application of pointers is discussed in Section 8-3.

Pointers to Arrays

Pointers can be aliased to array elements just as they are aliased to simple variables. Of course, the array must be defined with the target attribute. For example, the following declarations specify a real target array and a pointer to a real value:

```
real, target :: x(10)
real, pointer :: ptr
```

Assume that the values for the elements in the array have been initialized, and that we now want to determine the maximum value and print it. The following statements alias the pointer ptr to the maximum value and then print the maximum using the pointer alias:

```
ptr => x(1)
do k=2,10
   if (x(k) > ptr) ptr => x(k)
end do
print*, "Maximum value is ", ptr
```

If we assume that the maximum value is in x(3), then the corresponding memory snapshot is the following:

x(1)	−18
x(2)	27
x(3)	84 ◄── ptr
x(4)	36
x(5)	47
x(6)	24
x(7)	−3
x(8)	0
x(9)	75
x(10)	25

A pointer can also be defined with an entire array as its alias. For example, the following statements define an array of five integers, and a pointer to the entire array:

```
integer, target :: k(5)
integer, dimension(5), pointer :: ptr
```

If we assume that the array values have been defined, then the following statement assigns the ptr alias to the array and then prints all five values using the alias:

```
ptr => k
print*, ptr
```

Implementing Arrays of Pointers

An *array of pointers* cannot be defined in Fortran 90, but an array of pointers can be implemented using an array of derived data types in which the derived data type is a pointer. Consider the following definition:

```
type pointer_element
    integer, pointer :: ptr
end type pointer_element
```

Now consider the following array definition:

```
type(pointer_element) :: array(25)
```

This statement defines an array with elements that are pointers to integers.

The following statements define an array with five elements and then define an array of pointers. The pointers are initialized to point to the elements in the array.

```
type pointer_element
    integer, pointer :: ptr
end type pointer_element
integer :: k
integer, target :: x(5)
type(pointer_element) :: array(5)
...
do k=1,5
   x(k) = k - 5
   array(k)%ptr => x(k)
end do
```

The memory snapshot after these statements are executed is the following:

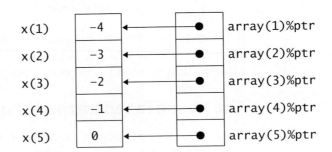

In Section 8-2 we will use arrays of pointers to sort information.

Pointers as Procedure Arguments

Pointers can be used as function arguments, but they are not listed on `intent` statements. However, if the procedure is an external procedure, then the calling program or procedure must include an *interface block* so that the compiler can generate a correct call to the procedure. The interface block contains only declarations of dummy arguments and result variables, and does not contain executable statements. For example, assume that we want a function to count the number of elements in an integer array that are greater than an integer that is aliased by a pointer. The function statements are the following:

```
!-------------------------------------------------------------!
integer function greater_count(x,npts,ptr)
!
!  This function counts the number of array elements
!  that are greater than a value aliased by ptr.
!
!    Declare variables.
     implicit none
     integer :: k, npts, x(npts)
     integer, pointer :: ptr
     intent (in) :: x, npts
!
!    Count values greater than ptr.
     greater_count = 0
     do k=1,npts
        if (x(k) > ptr) greater_count = greater_count + 1
     end do
!
!    Function return.
     return
!-------------------------------------------------------------!
```

To use this function as an external function, the following interface block needs to be used in the calling program or procedure:

```
interface
    function greater_count(x,npts,ptr)
        integer :: npts, x(npts)
        integer, pointer :: ptr
    end function greater_count
end interface
```

If the `greater_count` function is used as an internal function (see Section 8-3), the interface block is not needed because the compiler has all the information needed to correctly compile the program.

The associated Intrinsic Function

There are three possible situations or states for a pointer variable: It has not been assigned an alias (undefined), it has been assigned an alias *(associated)*, or it has been assigned a null value (nullified). When used with a single argument, the `associated` intrinsic function can be used to determine if a pointer variable has been associated with a variable. This function returns a value of true if the pointer has been associated and a value of false if it has been assigned a null value; this function should not be used with an undefined pointer. Thus, the expression `associated(ptr)` is true for the following memory snapshot:

while the expression `associated(ptr_1)` is false for this situation:

ptr_1

The `associated` function can also be used with two arguments, as in `associated(ptr,b)`. If the second argument `b` is a target variable, then the function is true if `ptr` is an alias for `b` and false if it is not. If the second argument is a pointer, as in `associated(ptr_1, ptr_2)`, then the function is true if `ptr_1` and `ptr_2` are either both null or if they are both aliases of the same target.

8-2 Application   ## MIGRATION PATHS

Environmental Engineering
In this problem we assume that the Wildlife Service has collected sighting information on the whooping crane migration. The data have been stored in a data file `cranes.dat`, in ascending order by the dates and times of the sightings. Each line in the data file corresponds to a sighting and contains the following four integer values:

- Sighting date (using year-day-month form, as in 950421)
- Sighting time (using a 24-hour clock, as in 1830)
- Grid location (an integer coordinate that gives location)
- Number of birds

Write a program that will read the data and then print sighting grid locations, both in the original order (by date and time) and in an ascending order (by grid location). Print the grid locations for the two orders in side-by-side columns.

1. Problem Statement

Write a program to print the grid locations of the current migration sightings of the whooping cranes in two orders: by date and by grid location.

2. Input/Output Description

Input—a data file containing the information on whooping crane sightings

Output—a report containing the grid locations ordered by date and by grid location

 ## 3. Hand Example

For a hand example, use the following sighting data:

Date	Time	Grid	Number of Birds
950403	0920	217	5
950405	0815	219	4
950408	0830	217	4
950501	1805	228	4
950510	0800	219	4
950510	0915	229	4
950515	0730	237	5

If we order this data by date and time, the migration report should contain the following information:

```
Whooping Crane Migration
Grid by Date   Grid by Location
217            217
219            217
217            219
228            219
219            228
229            229
237            237
```

 ## 4. Algorithm Development

We need to reorder the grid data, so we read the grid values into an array. Since the grid values are printed in side-by-side ordered columns, we need the data to be available in both orders at the same time. One solution would be to define a second array, sort the grid values into ascending order in that array, and then print the two arrays. Another solution is to define an array of pointers and then sort the pointers so that the first pointer points to the smallest grid value, the second pointer points to the next smallest grid value, and so on. We will use an array of pointers for the solution developed here.

A selection sort algorithm was presented in Chapter 5 and has been used in several programs in previous chapters. However, instead of reordering the data itself, we will reorder the pointers so that the first pointer will point to the data with the smallest grid value, the second pointer will point to the data with the next-largest grid value, and so on. Then, to print the data, we will use the original array to obtain the original order and the array of pointers to access the data in grid location order. To illustrate, using the hand example data, here are the value in the grid array and the corresponding array of pointers before the sort occurs:

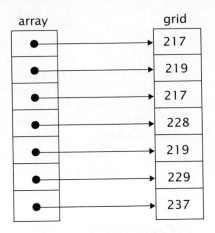

After the sort algorithm has been executed, the values in the pointer array should contain the following information:

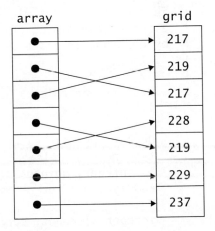

Decomposition

Read and store grid data in an array.
Sort the grid data by grid location.
Print the grid data ordered by date and ordered by location.

Pseudocode

```
migration_report:
      k ← 1
      while more data
            read grid(k)
            set array(k)%ptr to alias grid(k)
            add 1 to k
```

```
            n_sightings ← k − 1
            count ← n_sightings
            current ← 1
            while current ≤ count−1
                 next_min ← current
                 k ← current+1
                 while k ≤ count
                      if value aliased by array(k)%ptr <
                          value aliased by array(next_min)%ptr
                              next_min ← k
                      add 1 to k
                 switch values of array(current)%ptr and array(next_min)%ptr
                 add 1 to current
            k ← 1
            while k ≤ n_sightings
                 print grid(k) and value aliased by array(k)%ptr
                 add 1 to k
```

Fortran Program

```
!-----------------------------------------------------------!
program migration_report
!
!  This program reads migration data and prints
!  a report containing grid locations in order
!  by date and in order by grid location.
!
!
!    Declare variables.
     implicit none
!
!    Declare pointer element data type.
     type pointer_element
        integer, pointer :: ptr
     end type pointer_element
!
     integer :: count, current, k=1, next_min, n_sightings, &
                skip_date, skip_time
     integer, target :: grid(500)
     type(pointer_element) :: array(500), hold
!
!    Open file and read migration data.
     open(unit=9,file="cranes.dat",status="old")
     do
        read(9,*,end=15) skip_date, skip_time, grid(k)
        array(k)%ptr => grid(k)
        k = k + 1
     end do
  15 n_sightings = k - 1
!
!    Sort grid array using an array of pointers
!    and a selection sort algorithm.
     count = n_sightings
```

```
        do current=1,count-1
           next_min = current
           do k=current+1,count
              if (array(k)%ptr < array(next_min)%ptr) then
                 next_min = k
              end if
           end do
           hold%ptr => array(current)%ptr
           array(current)%ptr => array(next_min)%ptr
           array(next_min)%ptr => hold%ptr
        end do
   !
   !    Print grids in two orders, side-by-side.
        print*, "Whooping Crane Migration"
        print*, "Grid by Date    Grid by Location"
        do k=1,n_sightings
           print 20, grid(k), array(k)%ptr
   20      format(1x,i3,12x,i3)
        end do
   !
   !    Close file.
        close(unit=9)
   !
   end program migration_report
   !-------------------------------------------------------------------!
```

5. Testing

Using the data file from the hand example, the output is:

```
Whooping Crane Migration
Grid by Date    Grid by Location
217             217
219             217
217             219
228             219
219             228
229             229
237             237
```

8-3 LINKED LISTS

Pointer variables allow us to implement new types of data structures — *linked data structures*. While we focus specifically on a linked list, other linked data structures include stacks, queues, and trees. A *linked list* is similar to an array because it is a group of *nodes* that contain information and have an order: There is a first node, a second node, and so on. However, the linked list is different from an array in several ways. First, we do not use a subscript to reference a node in the linked list; instead, we use a pointer variable. Because the linked list is built by associating pointers to the nodes of the linked list as needed, we do not have to declare a maximum size for a linked list. Each node in the linked list has a pointer contained within it that points to the next node in the list. The following diagram of a node in a linked list shows the distinction between the data in the node and the pointer:

Linked list node:

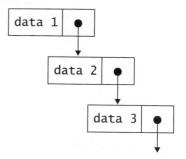

We can visualize the linked nodes in the following way, assuming that data 1 represents the data value in node 1, data 2 represents the data value in node 2, and so on:

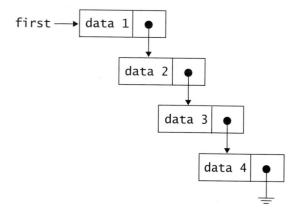

We need to be able to identify the first and last nodes in a linked list. To identify the first node, we assign a special pointer, usually called first, to point to the first node. We can then use a null pointer in the last node of a linked list to identify the end of the list. A diagram of a linked list with four nodes is illustrated in the next figure; note the use of a separate pointer to identify the beginning of the list and the use of the null pointer in the last node to identify the end of the list.

Initializing and Printing a Linked List

We now look at the steps necessary to initialize and store a given set of values in a linked list. We need three pointers to generate the list. One pointer, which we call first, aliases the first node in the list. (When the list is empty, first is a null pointer.) The other two pointers, previous and next, are used to point to the node most recently added to the list and to the next node to be added. The steps and resulting memory snapshots for generating a linked list are now described using these three pointers:

1. Allocate a node aliased by `first`.
   Move data into the node aliased by `first`.
   Alias the pointer `previous` to the first node.

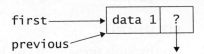

2. While there is more data to add to the list,
   allocate a node aliased by `next`,
   move new data into the node aliased by `next`,
   set the pointer in the previous node to alias the next node,

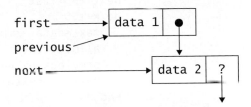

   set the previous pointer to alias the next node.

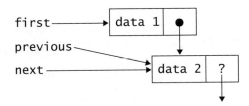

3. When there is no more data to add to the list, nullify the pointer stored in the last node.

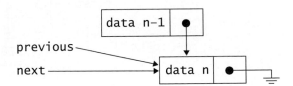

We now look at the Fortran 90 statements necessary to implement these steps that initialize values in this linked list. First we need to define a data type that contains a data value (which we will assume to be an integer value) and a pointer variable:

```
type node
    integer :: data_value
    type(node), pointer :: link
end type node
```

It is not necessary to define the nodes or elements in the linked list in the program declarations because the nodes of the linked list are allocated as needed.

EXAMPLE 8-1   ## Initializing a Linked List

Give the statements to generate a linked list that contains the data values 5, 10, 15, 20, and 25. Assume that data type `node` defined in the previous paragraph is available to the statements.

SOLUTION

Go through these steps by hand, drawing and updating the memory snapshot as this linked list is generated.

```
integer :: k
type(node), pointer :: first, previous, next
...
allocate(first)
next => first
do k=1,5
    next%data_value = k*5
    allocate(next%link)
    previous => next
    next => next%link
end do
nullify(previous%link)
```

The memory snapshot after the first pass through the loop is

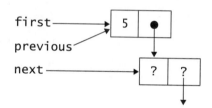

After all the statements are completed, the memory snapshot is

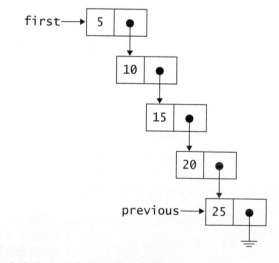

Now that we have covered the steps to build a linked list, we can use the links established to access the data. The next example illustrates the steps necessary to print the information stored in a linked list.

EXAMPLE 8-2   ## Printing the Data in a Linked List

Write a subroutine to print the data stored in the linked list generated in Example 8-1. Assume that the only argument to the subroutine is the pointer first that points to the first node in the list.

SOLUTION

To access and print the node in the linked list, we use the first pointer to point to the first node of the list, and a next pointer to determine and point to the next node in the list. If this subroutine is used as an external procedure, the derived data type definition for node should be included in an interface block in the main program; if the subroutine is used as an internal procedure, the interface block is not necessary.

```
!-----------------------------------------------------------------!
subroutine print_list(first)
!
!  This subroutine prints a linked list that
!  contains an integer value in each node.
!
!
!    Declare and initialize variables.
     implicit none
     type(node), pointer :: first, next
!
     if (.not.associated(first)) then
        print*, "Empty List"
     else
        print*, "Linked List Values"
        next => first
        do while(associated(next))
           print*, next%data_value
           next => next%link
        end do
     end if
!
!    Subroutine return.
     return
!
end subroutine print_list
!-----------------------------------------------------------------!
```

To be sure that you understand the steps in this subroutine, start with the memory snapshot of the list from Example 8-1, and follow through the steps in this subroutine that print the values in the linked list. The output should be

```
Linked List Values
5
10
15
20
25
```

Deleting Data in a Linked List

In addition to building and accessing the data in a linked list, we often want to delete old items or insert new items. These types of operations are especially efficient with linked lists. When we remove an item from a linked list, all we have to do is change the previous pointer so that it points to the item following the one we want to delete, as shown in the following diagram:

Before deleting the third node:

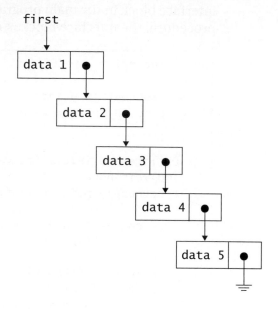

After deleting the third node:

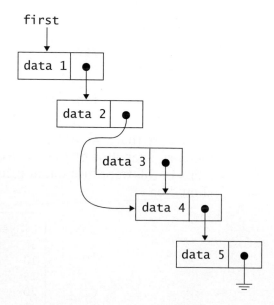

Even though the third node still points to the fourth node, the third node cannot be accessed because there is no pointer to it.

EXAMPLE 8-3   ## Deleting in a Linked List

Write a subroutine to delete a node from the linked list generated in Example 8-1. Assume that the arguments to the subroutine are the pointer first and the data value that is to be deleted. Since it is possible that there might be several nodes in the list with the same data value, the subroutine should delete only the first node with the data value. If the value is not in the list, print an error message.

SOLUTION

There are three cases to consider: deleting the first node in the list, deleting between two nodes in the list, and deleting the last node in the list. If we delete the first node, we only need to change the pointer first so that it points to the second node in the list, as shown in the diagram:

Before deleting the node at the front of the list:

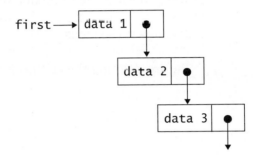

After deleting the node at the front of the list:

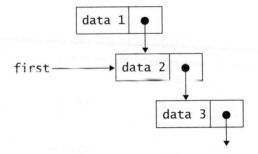

If we delete a node between two other nodes in the list, we modify the pointer in the preceding node so that it points to the node following the one we are deleting. This step is illustrated in the next pair of diagrams:

Before deleting a node between two nodes in the list:

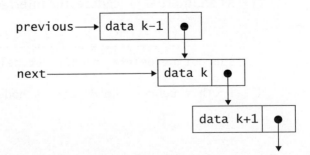

After deleting a node between two nodes in the list:

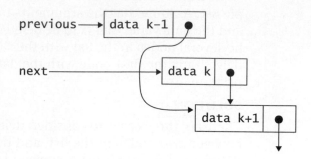

Finally, if we delete the last node, we modify the pointer in the previous node so that it contains a null pointer. The null pointer is contained in the pointer in the node to be deleted; thus, the steps here are the same as the steps for deleting a node between two nodes. The next pair of diagrams illustrates these steps:

Before deleting the last node in the list:

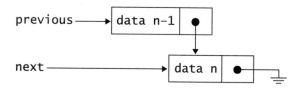

After deleting the last node in the list:

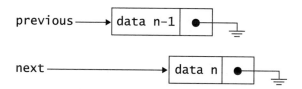

We now present the specified subroutine. If this subroutine is used as an external procedure, the derived data type definition for node should be included in an interface block in the main program; if the subroutine is used as an internal procedure, the interface block is not necessary.

```
!------------------------------------------------------------------!
subroutine delete_node(first,delete_value)
!
!  This subroutine deletes a node in a linked list.
!
!
```

```
!     Declare and initialize variables.
      implicit none
      integer :: delete_value
      logical :: done
      type(node), pointer :: first, next, previous
      intent (in) :: delete_value
!
      done = .false.
      if (.not.associated(first)) then
         print*, "Empty List"
      else
         if (first%data_value == delete_value) then
            first => first%link
         else
            next => first
            do while(.not.done)
               if (.not.associated(next)) then
                  done = .true.
                  print*, "Value not in Linked List"
               else
                  if (next%data_value > delete_value) then
                     done = .true.
                     print*, "Value not in Linked List"
                  else
                     if (next%data_value == delete_value) then
                        done = .true.
                        previous%link => next%link
                     else
                        previous => next
                        next => next%link
                     end if
                  end if
               end if
            end do
         end if
      end if
!
!     Subroutine return.
      return
!
end subroutine delete_node
!-----------------------------------------------------------------!
```

. . . .

Inserting Data in a Linked List

To insert a node in a linked list, we need to change two pointers. The pointer in the item before the new item has to be changed to point to the new item, and the pointer in the new node needs to point to the node that follows it. The following diagrams illustrate these steps:

Before insertion between second and third nodes:

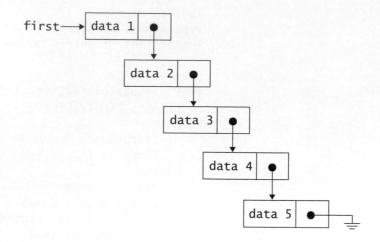

After insertion between second and third nodes:

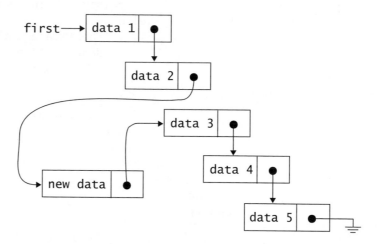

Recall that inserting an item in an array requires that all the items past the new item be moved one position down to make room for the new item. If the array is large, these steps can also be time consuming. With a linked list we only modify two pointers to insert an item.

EXAMPLE 8-4   Inserting in a Linked List

Write a subroutine to insert a node in the linked list generated in Example 8-1. Assume that the arguments to the subroutine are the pointer first and the data value that is to be added. We will insert the value so that the data values in the linked list are in ascending order. Since it is possible that there already is a node in the list with the same data value, the subroutine should insert the new node in front of any other nodes with the same data value.

SOLUTION

Again, we must consider three cases: inserting before the first node in the list, inserting between two nodes in the list, and inserting after the last node in the list. If we insert before the first node, we need to change the pointer `first` so that it points to the new node, and we need to change the pointer in the new node so that it aliases the node originally aliased by `first`. These steps are illustrated in the following diagrams, which assume that the pointer `current` aliases the new node that we want to insert:

Before inserting a new node at the front of the list:

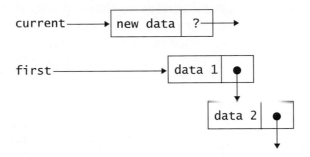

After inserting a new node at the front of the list:

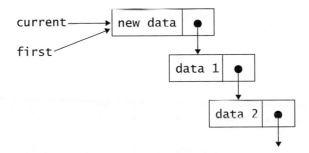

To insert a new node between two other nodes in the list that are pointed to by `previous` and `next`, we modify the pointer in the previous node so that it points to the new node, and then we modify the pointer in the new node to point to the next node. These steps are illustrated in the next pair of diagrams:

Before inserting a node between two nodes in the list:

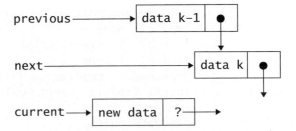

After inserting a node between two nodes in the list:

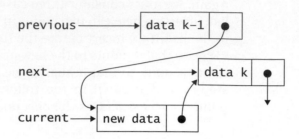

Finally, if we insert after the last node, we modify the pointer in the previous node so that it points to the new node, and then we modify the pointer in the new node to point to the null pointer. The null pointer is contained in the previous node, however, so the steps here are the same as the steps for inserting a node between two nodes, as shown below:

Before inserting after the last node in the list:

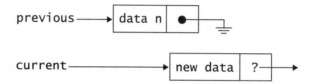

After inserting after the last node in the list:

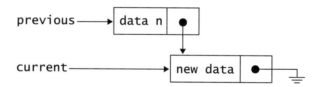

We now present the specified subroutine. If this subroutine is used as an external procedure, the derived data type definition for node should be included in an interface block in the main program.

```
!----------------------------------------------------------------!
subroutine insert_node(first,insert_value)
!
!  This subroutine inserts a new value in a linked list
!  that contains integer values in ascending order.
!
!
!    Declare and initialize variables.
     implicit none
     integer :: insert_value
     logical :: done
     type(node), pointer :: first, previous, current
     intent (in) :: insert_value
!
     allocate(current)
     current%data_value = insert_value
```

```
!
          if (.not.associated(first)) then
             first => current
             nullify(first%link)
          else
             if (first%data_value > insert_value) then
                current%link => first
                first => current
             else
                next => first%link
                previous => first
                done = .false.
                do while(.not.done)
                   if (.not.associated(next)) then
                      done = .true.
                      previous%link => current
                      nullify(current%link)
                   else
                      if (next%data_value > insert_value) then
                         done = .true.
                         previous%link => current
                         current%link => next
                      else
                         previous => next
                         next => next%link
                      end if
                   end if
                end do
             end if
          end if
!
!     Subroutine return.
      return
!
end subroutine insert_node
!----------------------------------------------------------------!
```

. . . .

EXAMPLE 8-5

Generating and Updating a Linked List

To test the subroutines for inserting and deleting a linked list, write a program to generate the linked list described in Example 8-1. Then allow the user to enter insertions and deletions to the list from the keyboard. Print the linked list before the updates and then again after all the updates have been completed. Use the subroutines developed in Examples 8-2, 8-3, and 8-4.

SOLUTION

```
!----------------------------------------------------------------!
program create_list
!
!  This program generates a linked list and then allows
!  the user to insert and delete values from the list.
```

```
!
!
!     Declare variables.
      implicit none
!
      type node
         integer :: data_value
         type(node), pointer :: link
      end type node
!
      integer :: code, k, update_value
      type(node), pointer :: first, next, previous
!
!     Generate initial list with values 5,10,15,20,25.
      allocate(first)
      next => first
      do k=1,5
         next%data_value = k*5
         allocate(next%link)
         previous => next
         next => next%link
      end do
      nullify(previous%link)
!
!     Print original values.
      print*, "Original Values"
      call print_list(first)
!
!     Allow user to update the list.
      print*, "To add a value, enter 1 followed by new value"
      print*, "To delete a value, enter 2 followed by old value"
      print*, "To quit, enter 9 9"
      read*, code, update_value
      do while (code /= 9)
         if (code == 1) then
            call insert_node(first,update_value)
         else
            call delete_node(first,update_value)
         end if
         read*, code, update_value
      end do
!
!     Print updated values.
      print*, "Updated Values"
      call print_list(first)
!
contains
!-------------------------------------------------------------------!
          (print_list subroutine from page 243 goes here)
!-------------------------------------------------------------------!
          (delete_node subroutine from page 246 goes here)
!-------------------------------------------------------------------!
          (insert_node subroutine from page 250 goes here)
```

```
!----------------------------------------------------------------------!
end program create_list
!----------------------------------------------------------------------!
```

We have included the subroutines as internal procedures and thus do not need to include interface blocks.

Here is the output from one test of this program:

```
Original Values
Linked List Values
5
10
15
20
25
To add a value, enter 1 followed by new value
To delete a value, enter 2 followed by old value
To quit, enter 9 9
1 8
2 10
2 25
1 28
2 17
Value not in Linked List
1 -3
1 3
1 7
2 3
9 9
Updated Values
Linked List Values
-3
5
7
8
15
20
28
```

. . . .

Summary

This chapter introduced the concept of a pointer, or an alias, to another variable. Examples were presented using pointers to variables, pointers to array elements, pointers as procedure arguments, and a technique for implementing an array of pointers. One of the most common uses of pointers is to implement a linked data structure. A program with three subroutines was developed to create a linked list, print a linked list, add nodes to a linked list, and delete nodes from a linked list.

Key Words

alias

array of pointers

associated

interface block

linked data structure

linked list

node

null pointer

pointer

Problems

This problem set begins with modifications to programs developed earlier in this chapter. Give the decomposition, refined pseudocode or flowchart, and Fortran solution for each problem.

Problems 1–5 modify the program migration_report given in Section 8-2 on page 238.

1. Modify the program so that the report lists the grids by date in descending order instead of ascending order. Keep the same ascending order for the grids by location.

2. Modify the program so that the report lists the grids by location in descending order instead of ascending order. Keep the same ascending order for the grids by date.

3. Modify the program so that a list of the grids by date is printed first, followed by a list of grids by location. Continue to use an array of pointers to reference the grids by location.

4. Modify the program so that it prints the grid location with most sightings. Also print the number of different sightings for this location.

5. Modify the program so that it prints the grid location with the fewest sightings. Also print the number of different sightings for this location.

Problems 6–10 modify the program create_list and its subroutines given in Section 8-3 on page 251 which generates a linked list.

6. Modify the delete_node routine so that it deletes all occurrences of the value to be deleted. (It is possible that several nodes contain the same value.)

7. Modify the insert_node routine so that it does not insert a node that contains the same value as a node that is already in the list. Print a message that the value is already in the node.

8. Write a real function average_node that computes the average value of the data values in the linked list pointed to by first. Assume that the function is referenced by

```
average_node(first)
```

9. Write a subroutine that determines the maximum and minimum values in the linked list pointed to by first. Assume that the subroutine is called with the reference

```
call link_min_max(first,min_value,max_value)
```

10. Write a real function that determines the standard deviation of the values in the linked list pointed to by `first`. Assume that the function is referenced by

$$\texttt{std\_dev(first)}$$

The next set of problems uses data entered from the terminal to display the status of orders at a computer warehouse. As people come by to pick up an item, the information is added to a linked list displayed on a screen in the warehouse so that the stock people can bring the items to the pick-up door. Once an order is picked up, the information is entered at the terminal so that the order can be removed from the list. Note that the items in this list are in the order in which they are received. Use the following derived data type definition in these problems:

```
type order_node
    integer :: order, item
    type(order_node), pointer :: link
end type order_node
```

The linked list is referenced by two pointers. The pointer `first` points to the first element in the list, and the pointer `last` points to the last element in the list. If `first` is a null pointer, the list is empty. If `first` and `last` point to the same element, the list contains one order.

11. Write a subroutine named `new_order` that is referenced with the following statement:

$$\texttt{call new\_order(order,item,first,last)}$$

The subroutine should add a node at the end of the list containing the integer order number and the integer item number.

12. Write a subroutine `print_orders` that is referenced with the following statement:

$$\texttt{call print\_orders(first)}$$

This subroutine should print all the information in the orders linked list. If the list is empty, print an appropriate message.

13. Write a subroutine `order_filled` that is referenced with the following statement:

$$\texttt{call order\_filled(first,order)}$$

The subroutine should delete the node in the linked list that corresponds to the order number. If the corresponding entry is not in the linked list, print an error message.

14. Write a program using the subroutines from problems 11, 12, and 13 to accept information from the terminal, either to add data to the end of

the linked list or to remove data from the linked list. Assume that the terminal interaction begins with this message:

```
Enter N to add new order to list
      D to delete order from list
      Q to quit program
```

If the character N is read, the program should print the message "Enter order number and item number." After reading the appropriate information, it should use the new_order subroutine to add the information to the list. If the character D is read, the program should print the message "Enter order number." After reading the order number, it should use the subroutine order_filled to remove the order from the list. If the character Q is read, the program should print either "Order List Empty" or "Order List Not Empty." If the list is not empty, the list should be printed before the program terminates. The program should be built around a main loop that accepts the character N, D, or Q from the terminal and then takes the appropriate action. After adding or deleting an order, the program should print the contents of the list and return to wait for the next character to be entered.

15. Modify the program in problem 14 so that it prints a final count of the orders filled before terminating.

Assume that an additional linked list is going to be used to store the item numbers and quantities sold for the computer warehouse. This new linked list is to contain the item number and the quantity of that item that has been sold, as indicated in this node definition:

```
type sold_node
    integer :: item, quantity
    type(sold_node), pointer :: link
end type sold_node
```

Each time an order is received, this linked list should be updated.

16. Write a subroutine that is referenced with the following statement:

```
call new_item(first_item,new_item)
```

The subroutine should insert a new item in the linked list, and give it an initial quantity sold of zero. Assume that the items are in ascending order and that first_item points to the first item in the list.

17. Write a subroutine print_item that is referenced with the following statement:

```
call print_item(first_item)
```

This subroutine should print the information in the items sold linked list. If the list is empty, print an appropriate message.

18. Modify the program in problem 14 so that it correctly updates the linked list of items sold when new orders are filled. Print a summary of the information in this linked list when the program terminates.

Appendix A

COMMON FORTRAN 90 INTRINSIC FUNCTIONS

In the following table of intrinsic functions, the names of the arguments specify their type, as indicated below:

Argument	Type
x,y	Real
chx	Character
dx,dy	Double precision
cx,cy	Complex, a + bi
ix,iy	Integer
gx,gy	Generic
px	Pointer
gax	Generic array
strx,stry	Character string

Function type, the second column of the table, specifies the type of value returned by an intrinsic function.

Generic function names are printed in bold. Any type argument that is applicable can be used with generic functions, and the function value returned is generally the same type as the input arguments, except for type conversion functions such as `real` and `int`.

Function Name	Function Type	Definition
sqrt(gx)	Same as gx	$\sqrt{gx}$
dsqrt(dx)	Double precision	$\sqrt{dx}$
csqrt(cx)	Complex	$\sqrt{cx}$
abs(gx)	Same as gx	$\lvert gx \rvert$
iabs(ix)	Integer	$\lvert ix \rvert$
dabs(dx)	Double precision	$\lvert dx \rvert$
cabs(cx)	Complex	$\lvert cx \rvert$

Function Name	Function Type	Definition
exp(gx)	Same as gx	e^{gx}
dexp(dx)	Double precision	e^{dx}
cexp(cx)	Complex	e^{cx}
log(gx)	Same as gx	$\log_e gx$
alog(x)	Real	$\log_e x$
dlog(dx)	Double precision	$\log_e dx$
clog(cx)	Complex	$\log_e cx$
log10(gx)	Same as gx	$\log_{10} gx$
alog10(x)	Real	$\log_{10} x$
dlog10(dx)	Double precision	$\log_{10} dx$
real(gx)	Real	Convert gx to real value
float(ix)	Real	Convert ix to real value
sngl(dx)	Real	Convert dx to single precision
anint(gx)	Same as gx	Round gx to nearest whole number
dnint(dx)	Double precision	Round dx to nearest whole number
nint(gx)	Integer	Round gx to nearest integer
idnint(dx)	Integer	Round dx to nearest integer
aint(gx)	Same as gx	Truncate gx to whole number
dint(dx)	Double precision	Truncate dx to whole number
int(gx)	Integer	Truncate gx to an integer
ifix(x)	Integer	Truncate x to an integer
idint(dx)	Integer	Truncate dx to an integer
ceiling(gx)	Integer	Least integer $\geq$ gx
floor(gx)	Integer	Greatest integer $\leq$ gx
sign(gx,gy)	Same as gx, gy	Transfer sign of gy to $|gx|$
isign(ix,iy)	Integer	Transfer sign of iy to $|ix|$
dsign(dx,dy)	Double precision	Transfer sign of dy to $|dx|$
mod(gx,gy)	Same as gx, gy	Remainder from gx/gy
amod(x,y)	Real	Remainder from x/y
dmod(dx,dy)	Double precision	Remainder from dx/dy
dim(gx,gy)	Same as gx, gy	gx − (minimum of gx and gy)
idim(ix,iy)	Integer	ix − (minimum of ix and iy)
ddim(dx,dy)	Double precision	dx − (minimum of dx and dy)
max(gx,gy,...)	Same as gx, gy,...	Maximum of (gx,gy,...)
max0(ix,iy,...)	Integer	Maximum of (ix,iy,...)
amax1(x,y,...)	Real	Maximum of (x,y,...)

Function Name	Function Type	Definition
dmax1(dx,dy,...)	Double precision	Maximum of (dx,dy,...)
amax0(ix,iy,...)	Real	Maximum of (ix,iy,...)
max1(x,y,...)	Integer	Maximum of (x,y,...)
min(gx,gy,...)	Same as gx, gy,...	Minimum of (gx,gy,...)
min0(ix,iy,...)	Integer	Minimum of (ix,iy,...)
amin1(x,y,...)	Real	Minimum of (x,y,...)
dmin1(dx,dy,...)	Double precision	Minimum of (dx,dy,...)
amin0(ix,iy,...)	Real	Minimum of (ix,iy,...)
min1(x,y,...)	Integer	Minimum of (x,y,...)
sin(gx)	Same as gx	Sine of gx, assumes radians
dsin(dx)	Double precision	Sine of dx, assumes radians
csin(cx)	Complex	Sine of cx
cos(gx)	Same as gx	Cosine of gx, assumes radians
dcos(dx)	Double precision	Cosine of dx, assumes radians
ccos(cx)	Complex	Cosine of cx
tan(gx)	Same as gx	Tangent of gx, assumes radians
dtan(dx)	Double precision	Tangent of dx, assumes radians
asin(gx)	Same as gx	Arcsine of gx
dasin(dx)	Double precision	Arcsine of dx
acos(gx)	Same as gx	Arccosine of gx
dacos(dx)	Double precision	Arccosine of dx
atan(gx)	Same as gx	Arctangent of gx
datan(dx)	Double precision	Arctangent of dx
atan2(gy,gx)	Same as gx, gy	Arctangent of gy/gx
datan2(dy,dx)	Double precision	Arctangent of dy/dx
sinh(gx)	Same as gx	Hyperbolic sine of gx
dsinh(dx)	Double precision	Hyperbolic sine of dx
cosh(gx)	Same as gx	Hyperbolic cosine of gx
dcosh(dx)	Double precision	Hyperbolic cosine of dx
tanh(gx)	Same as gx	Hyperbolic tangent of gx
dtanh(dx)	Double precision	Hyperbolic tangent of dx
dprod(x,y)	Double precision	Product of x and y
dble(gx)	Double precision	Convert gx to double precision
cmplx(gx)	Complex	gx + 0i
cmplx(gx,gy)	Complex	gx + gyi

Function Name	Function Type	Definition
aimag(cx)	Real	Imaginary part of cx
real(cx)	Real	Real part of cx
conjg(cx)	Complex	Conjugate of cx, $a - bi$
minval(gax)	Same as gax	Minimum value in array gax
maxval(gax)	Same as gax	Maximum value in array gax
product(gax)	Same as gax	Product of values in array gax
sum(gax)	Same as gax	Sum of values in array gax
len(strx)	Integer	Length of character string strx
len_trim(strx)	Integer	Length of strx without trailing blanks
trim(strx)	Character string	strx with trailing blanks removed
index(strx,stry)	Integer	Position of substring stry in string strx
char(ix)	Character	Character in the ixth position of collating sequence
ichar(chx)	Integer	Position of the character chx in the collating sequence
lge(strx,stry)	Logical	Value of (strx is lexically greater than or equal to stry)
lgt(strx,stry)	Logical	Value of (strx is lexically greater than stry)
lle(strx,stry)	Logical	Value of (strx is lexically less than or equal to stry)
llt(strx,stry)	Logical	Value of (strx is lexically less than stry)
associated(px)	Logical	True if px is associated, false if px is null pointer

Answers to Try It! Exercises

Page 14

1. invalid (decimal point)
2. valid
3. valid
4. valid
5. valid
6. valid
7. invalid characters (parentheses)
8. invalid (starts with a digit)
9. valid
10. invalid character (%)
11. same
12. not the same (− 1.7, 0.17)
13. same
14. same
15. not the same (0.899, 8990)
16. not the same (− 0.044, 0.044)

Page 21

1. y $\boxed{12.5}$

2. x $\boxed{4.1}$

3. result $\boxed{-13}$

Page 23

1. m = sqrt(x**2 + y**2)
2. u = (u + v)/(1.0 + (u*v/c**2))
3. y = y0*exp(-a*t)*cos(2.0*pi*f*t)
4. t = ((5.0/9.0)*(tf - 32.0)) + 273.15
5. $\text{potential energy} = \dfrac{-g \cdot me \cdot m}{r}$
6. $\text{flux} = e \cdot da \cdot \cos(\Theta)$
7. $\text{average velocity} = \dfrac{x_2 - x_1}{t_2 - t_1}$
8. $\text{centripetal acceleration} = \dfrac{4 \cdot \pi^2 r}{t^2}$
9. $\text{distance} = \text{velocity} \cdot \text{time} = \dfrac{\text{acceleration} \cdot \text{time}^2}{2}$
10. $\text{pressure} = p_0 \cdot e^{(-m \cdot g \cdot x \cdot tk/r)}$

Page 36

1. x$_b$=bbb-27.6$_b$degrees
2. distance$_b$=bb0.287E$_b$05bbbbbvelocity$_b$=b-2.60

Page 48

1. bb3.50bbbbbbbbbbbb*******b0.0020
2. time$_b$=bb3.50bbbbbbresponse$_b$1$_b$=bb0.18E+03
   time$_b$=bb3.50bbbbbbresponse$_b$2$_b$=bb0.20E-02
3. Experiment$_b$Results

   Time$_{bb}$Response$_b$1$_{bb}$Response$_b$2
   3.50$_{bbbbb}$178.800$_{bbbbbbb}$0.002
4. XX.XXX$_b$XX.XXX$_b$XX.XXX$_b$XX.XXX
5. XXX.XX
   XXX.XX
   XXX.XX
   XXX.XX
6. X.XX
   XX.X
   X.XX
   XX.X
7. XX.XXXXX.XXX
   XX.XXXXX.XX

Page 65

1. false
2. false
3. false
4. true

5. true
6. false
7. false
8. true
9. `if (time > 5.0) time = time + 0.5`
10. ```
    if (sqrt(polynomial) >= 8.0) then
        print*, "polynomial = ", polynomial
    end if
    ```
11. ```
    if (abs(volt_1 - volt_2) < 6.0) then
        print*, "volt_1 and volt_2:"
        print*, volt_1, volt_2
    end if
    ```
12. ```
    if (abs(denominator) < 0.005) then
        print*, "Denominator is too small"
    end if
    ```
13. ```
    if (log(x**2) >= 3.0) then
        time = 0
        sum = sum + x
    end if
    ```
14. ```
    if (distance < 50.0).or.(time > 10.0) then
        time = time + 1.0
    else
        time = time + 0.5
    end if
    ```
15. ```
    if (distance >= 50.0) then
        time = time + 2.0
        print*, "Distance > 50.0"
    end if
    ```
16. ```
    if (distance > 100.0) then
        time = time + 2.0
    else if (distance > 50.0) then
        time = time + 1.0
    else
        time = time + 0.5
    end if
    ```

Page 82

1. 10 times
2. 9 times
3. 13 times
4. 7 times
5. 9 times
6. 91 times
7. 8
8. 15
9. −16
10. 18
11. −1
    ```

Page 92

1. 33
2. 36
3. 5

Page 99

1. time [0.0]   temperature [86.3]

2. time [0.0]   temperature [0.5]

3. time_1 [0.0]   temperature_1 [86.3]
   time_2 [0.5]   temperature_2 [93.5]

4. time_1 [0.0]   temperature_1 [86.3]
   time_2 [0.5]   temperature_2 [93.5]

5. time_1 [0.0]   temperature_1 [0.5]
   time_2 [1.0]   temperature_2 [1.5]

6. time_1 [0.0]   temperature_1 [0.5]
   time_2 [86.3]   temperature_2 [93.5]

Page 106

1. works
2. works
3. would not work; time must be -99.0
4. would not work; the program will try to read a value for the variable altitude, causing an execution error
5. works

Page 125

1. array values:

```
m( 1) =    10
m( 2) =     9
m( 3) =     8
m( 4) =     7
m( 5) =     6
m( 6) =     5
m( 7) =     4
m( 8) =     3
m( 9) =     2
m(10) =     1
```

2. -3   0   5   12   21   32   45   60
3. time 1 = 3.00
    time 5 = 5.00
    time 9 = 7.00
    time 13 = 9.00
    time 17 = 11.00

Page 126

1. Sum = 17
2. 6720 is the product of elements
3. Maximum: 8
   Minimum: -4

Page 136

1.

31		8		8		8		8		8
24		24		16		16		16		16
63	→	63	→	63	→	21	→	21	→	21
16		16		24		24		24		24
21		21		21		63		63		31
8		31		31		31		31		63

Page 143

1.

4	6	8	10
6	8	10	12
8	10	12	14
10	12	14	16
12	14	16	18

   12   14   16   18

2.

1	2	0
2	3	1
3	4	2

   1   2   0
   2   3   1
   3   4   2

3.

11.5	14.5	17.5
13.0	16.0	19.0

   11.5 14.5 17.5
   13.0 16.0 19.0

Page 170

1. 5.8
2. 2.9
3. 5.3
4. 6.05

Page 184

1. New values of k are:

```
      1    2    3    0    1
      2    3    0    1    2
```

2. New values of k are:

```
      1    2    3    0     1
      2    3    0    9    10
```

Page 210

1. ten top engineering
2. ten top
3. engineering
4. a
5. engineering achievements
6. ten top
7. ten top engineering achievements
8. ten achievements
9. top ten
10. engineers' achievements
11. laser
12. fiber
13. CADCAM
14. """
15.   12.4
16. geneti

Page 220

1. 0.75D+00
2. 1.3D+00
3. 1.0D+00/9.0D+00
4. 5.0D+00/6.0D+00
5. -10.5D+00
6. 3.0D+00
7.   0.78692D-02
8.   0.787D-02
9. ************

Page 222

1. 2.0 + 6.0i
2. 1.0 + 0i
3. − 2.0i
4. 3.0 + 0.0i
5. 3.5 − 1.5i
6. 4.12 + 0.0i
7.   0.0 +   2.0 i
8. (0.00, 2.00)

Page 230

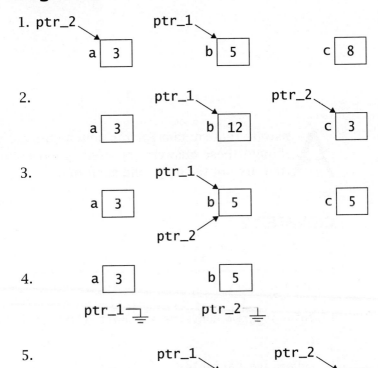

Answers to Selected Problems

A nswers that contain Fortran statements are not usually unique. Although these answers represent good solutions to the problems, they are not the only valid solutions.

CHAPTER 2

1.
```
!-----------------------------------------------------------------!
program convert
!
!   This program converts kilowatt-hours to calories.
!
!
!     Declare variables
      implicit none
      real :: calories, joules, kwh
!
!     Prompt the user to enter the energy value.
      print*, "Enter energy in kilowatt hours:"
      read*, kwh
!
!     Convert energy to joules and then to calories.
      joules = 3.6E+06*kwh
      calories = joules/4.19
!
!     Print kilowatt-hours and calories.
      print 5, kwh, calories
    5 format(1x,f6.2," kilowatt-hours = ",e9.2," calories")
!
end program convert
!-----------------------------------------------------------------!
```

CHAPTER 3

7.
```fortran
!----------------------------------------------------------------!
program table
!
!  This program prints a table of values for f where
!  f = xy - 1 for a given set of x and y values.
!
!
!     Declare and initialize values.
      implicit none
      integer :: i
      real :: f, x, y=0.5
!
!     Print table of values.
      print*, "x         y         f"
      do i=1,9
         x = real(i)
         f = x*y - 1.0
         print 5, x, y, f
    5    format(1x,f3.1,4x,f4.1,4x,f4.1)
         y = y + 0.25
      end do
!
end program table
!----------------------------------------------------------------!
```

CHAPTER 4

2.
```fortran
!----------------------------------------------------------------!
program sonar_signal
!
!   This program generates a sonar signal.
!
!
!     Declare and initialize variables.
      implicit none
      integer :: k, number
      real :: A, E, frequency, PD, pi, period, t, &
              time, signal
      parameter (pi=3.141593)
!
!     Open output file.
      open(unit=10,file="sonar1.dat",status="unknown")
!
!     Prompt user for input information.
      print*, "Enter transmitted energy in joules:"
      read*, E
      print*, "Enter pulse duration in seconds:"
      read*, PD
      print*, "Enter cosine frequency in Hertz:"
      read*, frequency
```

```
!
!      Compute signal parameters.
       A = sqrt(2.0*E/PD)        ! signal amplitude
       period = 1.0/frequency    ! signal period
       T = period/10.0           ! sample time interval
       number = nint(PD/T)       ! number of samples to compute
!
!      Generate signal values and write to a data file.
       do k=0,number-1
          time = real(k)*T
          signal = A*cos(2.0*pi*frequency*time)
          write(10,*) time, signal
       end do
!
!      Write trailer signals and close file.
       time = -99.0
       signal = -99.0
       write(10,*) time, signal
       close(unit=10)
!
end program sonar_signal
!-----------------------------------------------------------------!
```

CHAPTER 5

3.
```
!-----------------------------------------------------------------!
program power_plant
!
!   This program computes and prints a composite report
!   summarizing several weeks of power plant data.
!
!
!      Declare and initialize variables.
       implicit none
       integer :: count=0, i, j, minimum, n, power(8,7), sum=0
       real :: average
!
!      Read the number of weeks to use for the report.
       print*, "Enter number of weeks (maximum 8) for report:"
       read*, n
!
!      Open file and read data into the array.
       open(unit=12,file="plant.dat",status="old")
       do i=1,n
          read(12,*) (power(i,j),j=1,7)
       end do
!
!      Compute minimum power and average power. Then determine
```

```
!      number of days with power greater than the average.
       minimum = power(1,1)
       do i=1,n
          do j=1,7
             if (power(i,j) < minimum) minimum = power(i,j)
             sum = sum + power(i,j)
          end do
       end do
       average = sum/real(7*n)
       do i=1,n
          do j=1,7
             if (power(i,j) > average) count = count + 1
          end do
       end do
!
!      Print initial part of the report.
       print 30, average, count
   30 format(1x,15x,"Composite Information"// &
            1x,"Average Daily Power Output = ", &
               f5.1," megawatts"// &
            1x,"Number of Days with Greater Than ", &
               "Average Power Output = ",i2/)
!
!      Determine and print days with minimum power output.
       print 35
   35 format(1x,"Day(s) with Minimum Power Output:")
       do i=1,n
          do j=1,7
             if (power(i,j) == minimum) print 40, i, j
   40        format(1x,7x,"Week ",i2,"    Day ",i2)
          end do
       end do
!
!      Close file.
       close(unit=12)
!
end program power_plant
!-----------------------------------------------------------------!
```

CHAPTER 6

11.
```
!-----------------------------------------------------------------!
logical function above_zero(x,npts)
!
!   This function determines whether or not all the
!   values in the array x are above zero. If they are
!   above zero, the function returns a true value;
!   otherwise it returns a false value.
!
!
```

```
!     Declare variables.
      implicit none
      integer :: k, npts
      real :: x(npts)
      intent (in) :: x, npts
!
!     Check for values above zero.
      above_zero = .true.
      for k=1,npts
          if (x(k) <= 0.0) above_zero = .false.
      end do
!
!     Function return.
      return
!
end function above_zero
!------------------------------------------------------------------!
```

CHAPTER 7

9.
```
!------------------------------------------------------------------!
subroutine alphabetize(letters)
!
!  This subroutine alphabetizes the letters
!  in an array containing 50 characters.
!
!
!     Declare variables.
      implicit none
      integer :: count, current, k, next_min
      character*1 :: hold, letters(50)
      intent (inout) :: letters
!
!     Use a selection sort algorithm to alphabetize letters.
      count = 50
      do current=1,count-1
         next_min = current
!
!        Find next minimum.
         do k=current+1,count
            if (letters(k) < letters(next_min)) next_min = k
         end do
!
!        Switch the current position with next minimum.
         hold = letters(current)
         letters(current) = letters(next_min)
         letters(next_min) = hold
      end do
!
!     Subroutine return.
      return
!
end subroutine alphabetize
!------------------------------------------------------------------!
```

CHAPTER 8

9.
```
!-------------------------------------------------------------!
program migration_report
!
!   This program reads migration data and prints
!   a report containing grid locations in order
!   by date and in order by grid location.
!
!
!      Declare variables.
       implicit none
!
!      Declare pointer element data type.
       type pointer_element
           integer, pointer :: ptr
       end type pointer_element
!
       integer :: count, current, k=1, max_count, max_grid, &
                  n_sightings, next_min, old_count, old_grid, &
                  skip_date, skip_time
       integer, target :: grid(500)
       type(pointer_element) :: array(500), hold
!
!      Open file and read migration data.
       open(unit=9,file="cranes.dat",status="old")
       do
          read(9,*,end=15) skip_date, skip_time, grid(k)
          array(k)%ptr => grid(k)
          k = k + 1
       end do
   15  n_sightings = k - 1
!
!      Sort grid array using an array of pointers
!      and a selection sort algorithm.
       count = n_sightings
       do current=1,count-1
          next_min = current
          do k=current+1,count
             if (array(k)%ptr < array(next_min)%ptr) then
                next_min = k
             end if
          end do
          hold%ptr => array(current)%ptr
          array(current)%ptr => array(next_min)%ptr
          array(next_min)%ptr => hold%ptr
       end do
!
!      Print grids in two orders, side-by-side
!      and find grid with most sightings.
       old_grid = array(1)%ptr
       old_count = 0
       max_grid = array(1)%ptr
```

```
        max_count = 0
        print*, "Whooping Crane Migration"
        print*, "Grid by Date    Grid by Location"
        do k=1,n_sightings
           print 20, grid(k), array(k)%ptr
  20       format(1x,i3,12x,i3)
           if (old_grid == array(k)%ptr) then
              old_count = old_count + 1
           else
              if (old_count > max_count) then
                 max_grid = old_grid
                 max_count = old_count
              end if
              old_grid = array(k)%ptr
              old_count = 0
           end if
        end do
        if (old_count > max_count) then
           max_grid = old_grid
           max_count = old_count
        end if
!
!       Print grid with most sightings
        print*
        print*,"Grid with maximum sightings:"
        print 25, max_grid, max_count
  25 format(1x,"Grid ",i5," had ",i5," sightings")
!
end program migration_report
!-----------------------------------------------------------------!
```

Index